Vacuum Bubbling

Vacuum Bubbling introduces the background and applications for generating bubbles under a vacuum condition, accomplished through depressurization without the need to heat water. It presents the advantage of utilizing vapor bubble in deaeration applications because the diffusion for degassing happens between the water body and micro vapor bubbles without the need of membrane or packing.

Instead of relying on massive heating, vacuum bubbling focuses on depressurization down to the level of saturated vapor pressure or below to secure vapor bubbles with virtually zero dissolved non-condensable gases, including oxygen. The book considers prospective applications, such as extracting high-oxygen-content air from water for underwater breathing, pretreatment of aircraft fuel before being pumped into a fuel tank system, and probable desalination applications through massive bubbling combined with low-grade renewable energy.

The book is intended for researchers in thermal fluids, heat and mass transfer, process engineering, and water treatment fields and industry professionals working in power generation, plant and process engineering, transportation, and energy.

Yong-Du Jun joined the Faculty of Mechanical and Automotive Engineering at Kongju National University, Korea, in 1997, after receiving his doctoral degree in aerospace engineering from the University of Cincinnati. Prior to joining the faculty position at KNU, he worked in the field of propulsion system, engine aerodynamics, and turbomachinery erosion with CFD. His research at KNU focused on waste heat recovery, fluidized bed heat exchanger, heat transfer enhancement, and fouling reduction while longtime advising senior capstone design projects related to innovative utilization of renewables. He has a strong interest in the study of "vacuum bubbling" and the potential to minimize unnecessary energy use. Dr. Jun served as the chairman of the Division of Mechanical and Automotive Engineering and the dean of Academic Affairs of the Kongju National University.

Vacuum Bubbling
Techniques and Applications

Yong-Du Jun

CRC Press
Taylor & Francis Group
Boca Raton London New York

CRC Press is an imprint of the
Taylor & Francis Group, an **informa** business

First edition published 2024
by CRC Press
2385 NW Executive Center Drive, Suite 320, Boca Raton FL 33431

and by CRC Press
4 Park Square, Milton Park, Abingdon, Oxon, OX14 4RN

CRC Press is an imprint of Taylor & Francis Group, LLC

ISBN: 978-1-032-44936-4 (hbk)
ISBN: 978-1-032-44937-1 (pbk)
ISBN: 978-1-003-37462-6 (ebk)

DOI: 10.1201/9781003374626

Typeset in Times
by Apex CoVantage, LLC

Contents

Preface

The subject of this book deals with humanity's efforts to alleviate climate disaster, an unprecedented crisis for civilization. The Intergovernmental Panel on Climate Change's (IPCC) critical awareness of climate change and its ripple effects and the International Renewable Energy Agency's (IRENA) agenda for a sustainable future are realistically difficult to guarantee success due to lack of implementation capabilities, but at least the presentation of such a direction will be remembered by the human community living in this era, including me, and by subsequent generations as a message about how this generation, facing a crisis of civilization, was conscious of the situation and sought to improve it. Although it is not very feasible, I am grateful to those involved in this effort who present a vision for it.

Approaches to analyzing climate change problems and making them sustainable have been presented through materials such as IRENA's World Energy Transitions Outlook 2022, but the overall report card is not so promising. Among them, the industry's energy efficiency field was designated as one of the most important fields, accounting for 25% of the total carbon dioxide reduction target, but its performance was particularly poor. One of the reasons may be that, given the nature of the development of human civilization, the driving force behind value creation in general industries is making money, not improving the environment. In other words, from the perspective of industrial entities, the agenda of improving energy efficiency is no one's responsibility and corresponds to an ownerless drive. Would it be too much to say that the incentives provided by the government are just a pretense of being pushed? Another barrier is that no one in the field can deny the current process, so it is not easy for innovative ideas to come from within the industry. And lastly, there are not many technological alternatives to improve energy efficiency.

Having worked as a professor with no experience in industry, I am relatively free from these realistic constraints. With an academic background in mechanical engineering, I have been thinking about how much I can be involved and what my limitations are in dealing with the topic of this book, the vacuum phenomenon, the phase change of materials, and their application to industry. These concerns apply to everyone there may not be many people for clues to the problem in this big picture.

The new term *vacuum bubbling* was chosen because it is an expression that even non-experts can understand without much difficulty. It refers to the application of creating bubbles in a vacuum. The reason that books in this field are necessary is that specialized industrial fields such as vacuum processing are already fully commercialized technology areas, and it has become difficult to find reference materials where the technology has been patented and published. For example, most energy-saving vacuum degassing technologies are registered as patents for companies and are only used for promotional purposes to sell products, and academic content related to the technology behind them is not publicly introduced or discussed. In contrast, this book contains relatively detailed content that can help you understand the phenomenon of vacuum bubbling. Among the reasons for disclosing this information through the book, the most important thing is that the technology using the principles introduced is to be systematically studied by more people so that the development trajectory of human civilization, which has been driven by high carbon, high temperature, and high pressure, can be changed in a slightly sustainable direction. This is because it takes a lot of effort to change it.

Among the sayings of the old Chinese sage Mencius, there is a saying called "何必曰利." This is an answer given by Mencius in a conversation with the king of a country during the Warring States period in China after hearing the king talk about various plans, including territorial expansion, to benefit the country. His point is, Why does the king only talk about benefits? In order to lead the people well, other than money, it is also important to create a good social atmosphere, such as trust and etiquette. As many of you would expect, his claims were not recognized at that time, but many people still remember them 2,300 years later.

I am not an expert in this field. However, I am happy to be able to introduce a field of technology that can contribute to creating a better society by applying technology within my understanding. If the research results on this topic had been well introduced and established by seniors or fellow researchers, I would not have had to carry out this work. However, since this was not the case, I took on the challenge despite my lack of capacity and awareness of the problem. Through the challenge process, I was able to taste many of the fruits of the academic fields that senior researchers have achieved so far, and thanks to such verified information, this trip went relatively smoothly. In particular, the results of phase change research in the field of thermal engineering could be applied as the principle of room-temperature vapor bubble creation, and the behavior characteristics of objects in the metastable region were thought to be the Creator's gift to humanity. This book introduces four fields that can be applied through the vacuum bubbling methodology. Out of these, the industrial degassing field, was explained in relative detail, and the results are considered encouraging. Among other application areas, desalination appears to be a very challenging but also open field of application. In any case, if this book is widely introduced to interested students, researchers, technology developers, and policy-makers so that the prepared technology can blossom and become a technology helpful to humanity, it would be a great comfort and joy to me, who has taken "the road not taken" by others to take on a seemingly reckless challenge.

If this book is successfully published, the biggest event leading up to it will date back to the late spring of 2022, when Kyra Lindholm, mechanical editor at CRC Press, expressed her interest in an academic presentation I presented. I would like to express my gratitude and respect to her insight and leadership in seeing the value of this topic, which would otherwise have remained an old note in a box. I would also like to thank my research colleagues, Dr. Jong-Soo Lee; brothers Jin-Koo Kim and Je-Koo Kim, who sympathized with the importance of this field and were always with me; and CEO Seung-Won Jeon of JNT, who supported the experimental equipment. I would like to thank Professor Kwang J. Kim of UNLV and Dr. Jaedal Lee of POSCO for their positive comments on the writing of this book. I would like to thank my graduate student Hong Jeong-A and undergraduate senior Won-Jae Lee for their help with the experiment, and I would also like to express my gratitude to Hyun-Soo Kim and Yu-Been Lee, who are currently under study on the topic for their undergraduate theses, along with the encouragement. I would like to thank Ms. Sonia Tam of CRC Press for her hard work in bringing this rough and inadequate manuscript to light. I would also like to thank the many individuals, publishers, associations, and companies who willingly allowed us to cooperate in copyright use during the preparation of the manuscript. I would also like to express my gratitude and respect to the leadership of Google for providing free Google translate and Google Bard, which were of great help in writing the manuscript.

Lastly, I present this book to my beloved wife and two sons, who have been diligently practicing surfing the world's waves with me, as unknown others also do.

Yong-Du Jun
At Kongju University Cheonan Campus

1 Introduction

In early 2023, I submitted an article [1] titled "Energy field challenges to respond to climate change – focusing on IRENA's World Energy Transitions Outlook 2022" in the newsletter of the *Korean Society for New and Renewable Energy*. The main point of the article is that Korea and other IPCC convention countries have presented global warming reduction goals [2] to limit global climate change as much as possible, and to achieve this, each country has announced plans related to carbon neutrality in consideration of their respective field situations. However, in the Korean government's plan related to carbon neutrality, the industry reduction target is reduced from 14.5% to 11.4%, allowing an increase in emissions of 8 million tons, and instead, the proportion of nuclear power generation is increased from 23.9% to 32.4% to further reduce 4 million tons, and the remaining 4 million tons would be reduced in the overseas reduction sector and carbon capture technology (CCUS). In addition, the share of renewable energy generation has actually decreased from 30.2% to 21.6% as of 2021. In this regard, environmental groups protested that it was a declaration of abandonment of response to the climate crisis, and argued that the achievement of the goal could not be guaranteed, considering the current level of technological development, such as carbon capture [3]. Looking at the background of the relaxation of the industrial sector reduction target, the government's reduction target two years ago was 14.5%, but now the expected performance based on the opinion of the industry is only 5%. On this issue, the Ministry of Industry and the Carbon Neutral Green Growth Committee found a compromise of 11.4% through discussion. It is a very worrisome reality that a country with a carbon reduction budget of about 90 trillion won over the next five years jumps from 14.5% to 5% and 11.4% of the reduction target.

Then, what is the situation of international efforts to respond to global warming? IPCC, an intergovernmental on climate change under the United Nations, strongly warns that the fate of the Earth depends on it in the next ten years, saying that there are no more options in the climate crisis through the sixth comprehensive report on climate change. While presenting a goal of reducing greenhouse gas emissions by 43% compared to 2019 by 2030, it was proposed to reduce fossil fuels and expand to renewable energy. In addition, the importance of so-called "climate-resilient development," which will achieve sustainable development while reducing carbon emissions along with the use of cutting-edge technologies such as carbon capture, was also emphasized. Through the IPCC Sixth Comprehensive Report, which contains an unprecedentedly strong warning, the IPCC stated that the international community must act now to achieve the enhanced reduction target. IPCC chairman Hoesung Lee emphasized that the key to the sixth report is to solve the climate crisis and that the climate crisis is now a real problem directly related to national economy, and urged countries to take immediate action.

What is the achievement of the international community's efforts to respond to climate change so far? The answer can be found in the "World Energy Outlook 2022,"

DOI: 10.1201/9781003374626-1

published in 2022 by the International Renewable Energy Agency (IRENA) [4]. First, IRENA designates electrification and efficiency as key drivers of a possible energy transition through renewables, hydrogen, and sustainable biomass to realize a 1.5°C pathway, aiming to reduce CO_2 emissions by nearly 37 gigatons per year by 2050. And this reduction target will be achieved in six ways: (1) significant increases in renewable-based electricity generation and direct use, (2) substantial improvements in energy efficiency, (3) electrification of end-use sectors (e.g., electricity vehicles and heat pumps), (4) clean hydrogen and its derivatives, (5) bioenergy combined with carbon capture, and (6) last mile uses of carbon capture and storage. The interesting part is that renewable energy and energy efficiency improvement were considered the most important, with 25% each, and electrification, hydrogen, and carbon dioxide capture are expected to play a role in reducing emissions in that order.

What about the achievement? IRENA's official position on overall achievement is that despite some progress, the energy transition is not on track and that sufficient action is to be taken by 2030 to meet the newly pulled climate target for 2040 under the IPCC Sixth Report in 2023.

Scores are presented for each responsible sector; however, let's take a look at the subject of this book, the energy efficiency. First, the improvement rate by energy source is 1.2%/yr compared to the target of 2.9%/yr and shows an achievement of 41%, while the investment demand for energy efficiency of 1.5 trillion US$/yr is fulfilled by 0.3 trillion US$/yr, showing an achievement of 20%. So what was the reason for the lack of progress? Similar to other fields, it is understood that issues in the field of energy efficiency improvement seem not simple. This is because all energy technologies and facilities are actually being used in the field, and their replacement is expected to cause significant socio-economic resistance, along with huge replacement costs and fundamental changes to the landscape in some of the existing industries. Second, there are not many available options for candidate technologies for improving energy efficiency significantly. The entity that formulates policies can only provide administrative support but cannot provide new technologies. The most experts in their technology fields are those who rely on existing technology. In such an environment, it is certain that even if innovative ideas and technologies are proposed, it is not easy for the existing technology community to accept them, even if the technology itself is excellent. As a result, an innovative technology for energy efficiency improvement can hardly survive to be applied in a real field. In this context, I thought that clarifying the academic basis first before approaching the subject of vacuum bubbling industrially would be the way for the proposed technology to gradually settle down while minimizing side effects from a long-term perspective. Shall we start talking about the subject of vacuum bubbling?

Albeit bubbling is so common in everyday life and also in numerable industries, most of the bubble studies in mechanical applications are focused on the bubble generation in the process of boiling heat transfer point of view. My story in this book is oriented to bubbling in water, especially under a vacuum condition which did not get due attention from the conventional engineering practices, at least before the fossil energy depletion and the global warming issues hit all of us. The motivation of writing this book came from a long line of thinking as one of the contemporary engineers who have a feeling of some responsibility for all this generation has built for mankind but left unwelcomed trace for Earth and for generations to visit the planet Earth.

1.1   RAISING PROBLEMATIC ISSUES IN OUR WAY OF ENERGY USE

Starting with James Watt's steam engine, to all internal combustion engines and power plants, with the exception of solar power, which produces energy directly in the form of electricity, in many engineering applications in the industrial age, people rely on phase changes of matter to get mechanical power. The use of this phase change phenomenon seems indispensable for power generation until the existing energy sources remain available, but it is time to reflect on whether all the ways we use energy in our daily lives are appropriate or not. In general, at least in many engineering applications, high-temperature and high-pressure driving conditions have been used extensively in the pursuit of mass production and mass consumption without paying attention to how much energy goes into that process, as is the case with conventional transportation vehicles. In the case of conventional transportation vehicles, we all have agreed to use cars with 30% thermal efficiency for over a century. What about fancy gas turbine engines that bring us from Seoul to New York? When we burn it, it provides enough mechanical power to move anything we want! Do you think it is to be continued? Till when?

1.2   GIVING AN EXAMPLE OF INADEQUATE ENERGY USE: THE USE OF COMPRESSED AIR TO OBTAIN VACUUM

A trivial example of our spoiled use of energy I came across during my study of vacuum bubbling was the use of vacuum generator, a device that is connected to an air compressor tank to vacuum out air inside the water vessel. With it, in order to generate a vacuum condition, we first compress air and save it in a high-pressure air tank and release it through the vacuum generator, which is basically passing out the compressed air through a nozzle with a side hole. Through this process, the air inside the container escapes, generating a vacuum condition inside the container. The working principle of this vacuum generator is as follows: (1) compress air using a 5 hP air compressor; then (2) release air through the vacuum generator; (3) then the local pressure inside the vacuum generator becomes lowered at the downstream of the ejector nozzle; and (4) the air in the water vessel flows to the lowered pressure port and is kicked out of the tank, leaving a vacuumed space inside the vessel. Even in this process of vacuuming, what we are doing is that we consume energy in the process of compressing the air that is eventually to be thrown out. Vacuum is obtained as a by-product of throwing out high pressurized air. What we are actually doing is that in order to have a vacuum, we resort to high-pressure air, which does not make a good sense as a way to achieve vacuum, though it is considered cheap. Then, what else? Likewise, we see industrial processes that seem not the best or adequate way of consuming limited energy sources. One not trivial example can be a degassing or deaeration process of liquid water.

1.3   GIVING A MORE SERIOUS AND SUBJECT-RELATED EXAMPLE OF ENERGY MISUSE

Let's take a look at one of the major industrial applications, for example, a deaeration practice. Deaerators are extensively used in various industries, such as power generation, including nuclear, manufacturing processes of food and beverage, pharmaceutical, cosmetics, pulp and paper, and semiconductor as well. According to several market analysis reports, the deaerator machine market value worldwide is of about 4 billion dollars and is expected to grow at a compound annual growth rate (CAGR) of 4.5% till 2028 [5]. One of the most widely used deaerator types in the power industry is called *thermal deaerator*, in which feedwater is preheated and sprayed into a vessel filled with steam. Inside of the vessel, water droplets are supposedly further being heated by separately supplied steam and evaporated, through which the non-condensable gases are separated from water vapor and eventually vented out. Vapor inside the vessel is condensed to water but still at high temperature, which, they say, is not a big deal for boiler applications, in which this water is supposed to be heated again to generate steam anyway. However, in this process, typically a large vessel which contains a tray structure, injection facilities, and a separate boiler system to supply steam make the system bulky and large, claiming a significant amount of initial installation cost. On top of that, the energy required for pressurization, heating, and evaporating seems huge.

1.4 UNDERSTANDING FEATURES AND THE WORKING PRINCIPLES OF CONVENTIONAL DEAERATOR

The power deaerator, which is one of the deaerators operated by the existing thermal method introduced earlier, is widely used for power plants that drive steam turbines or for district heating, and its operation method can be roughly seen as follows: First, the feedwater is heated and then sprayed through a spray nozzle into the tank in a hot steam environment. Raising the temperature of water may ultimately be a requirement at the point of use of the supply water, but since the solubility decreases with temperature (Figure 1.1), it provides conditions in which dissolved gases in water can be separated from droplets more easily.

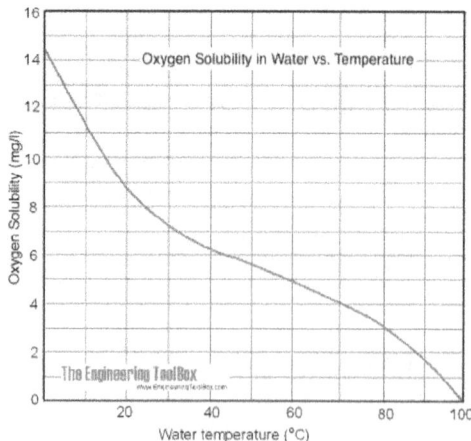

FIGURE 1.1   Solubility of oxygen in water [6].

The sprayed liquid droplet is further heated by the surrounding steam and vaporized, through which diffusion and evaporation proceed, and eventually the non-condensable gases dissolved in the water are separated, and the separated steam becomes degassed water through a condensation process. In the meantime, non-condensable gas and some steam are discharged to the outside through a vent. In order to increase the degassing level of the deaerated water collected in the container, the second deaeration by steam is performed through a steam sparger in the water. This series of processes is a method using mass transfer by steam and phase change, accompanied under a heated and pressurized condition. This process proceeds in the upper right corner (under high temperature and high pressure) of the phase diagram for water (Figure 1.2) and also requires equipment for pressurization and heating and consumption of energy. However, in terms of the physical mechanism related to degassing, it can be said that mass transfer is realized through diffusion by steam bubbling with very low oxygen concentration after discharging excessively dissolved solutes in advance by bringing water into a low saturated solubility state. Realizing a lower solubility state to discharge supersaturated solutes and mass transfer by high-temperature vapor (steam) bubbling can also be realized simply by reducing pressure without resorting to heating, pressurization, and external steam supplied separately. A properly designed device can have advantages in terms of device structure, size, energy cost, and operability.

Another popular deaerator type is membrane type, which is extensively used in the food and beverage and pharmaceutical fields. The main principle of the membrane separation process is diffusion. This application is considered to be heavily relying not only on the temperature field but also on the pressure field to obtain reasonable output rates. Further, this system has a critical weak point involved with the failure of membrane as well as the required regular change of the membrane, causing operational charges as well as energy charges. In a vacuum deaerator, water flows by

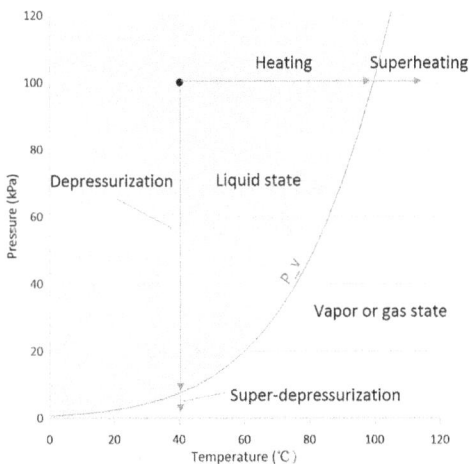

FIGURE 1.2 Phase diagram of water that illustrating two different approaches, heating and vacuuming, to reach a vaporization condition.

gravity down through a tower filled with packing as a vacuum is drawn on the tower. The packing in the tower has a very high surface area and disperses the water very effectively, thereby enhancing the removal of O_2, CO_2, and N_2.

1.5   WHAT IS VACUUM BUBBLING?

Vacuum bubbling is a newly coined name for a process of generating intrinsic bubbles out of liquids under a vacuum condition. *Intrinsic bubbling* refers to a phenomena of bubble generation that does not rely on any external gas sources like air, nitrogen, helium, etc. but comes out of the liquid itself, which may contain dissolved gas in it along with the vapor of liquid. Among these, the generation of bubbles due to changes in saturation solubility is directly driven by the solubility limit, which is known to be a function of pressure and temperature. Everyone experiences this phenomena in daily life, such as in sparkling of CO_2 bubbles from carbonated beverages. However, the generation of bubbles that occur when a liquid vaporizes is related to the phase change of a part of the liquid to a vapor state due to a change in the local thermodynamic state inside the bulk liquid.

Dissolved gas in liquids, and its treatment, has long been an interesting issue from many and diversified applications, such as "artificial gills" [7–10], deaeration processes in the food and beverage industry [11, 12], energy industry [13, 14], semiconductor manufacturing industries, etc. The efforts to develop artificial gills include studies to develop underwater breathing systems that can directly extract and separate dissolved oxygen (DO) from water, most approaches of which rely on polymeric membrane technologies or hollow fibers to obtain DO from pressurized water, as noted by Lee et al. [7]. One of the key drawbacks of this pressure-driven membrane technology is that the process requires significant amount of external energy mainly for pressurization. *Deaeration* refers to the removal of dissolved gases, such as oxygen, from liquids. Pressure deaerators, known to be the most efficient and are used in all power plants, rely their operation on the diffusion of gas between oxygen-free steam and sprayed droplets, in which steam not only contributes to the diffusion of gas between the two phases (liquid-phase droplets to gas-phase steam) but also heats the fine water droplets up near to saturation temperature to help evaporation [15]. Further, typical power deaeration systems are composed of highly pressurized feedwater inlet, a deaerator vessel with necessary arrangements, steam supply, and feed pump, which make the system bulky and complicated, consuming significant energy.

Now, noting that the key mechanisms involved with the current degassing process include diffusion between liquid-phase droplets and vapor-phase steam under a pressurized and heated condition resulting in high-energy consumption, I propose a low-pressure operation of the equivalent process as a low-cost, low-energy-consuming countermeasure to accomplish the same goal [16]. The working principle of conventional deaerators is mainly based on the diffusion of DO in the droplets to virtually oxygen-free steam with additional temperature-driven heating and evaporation, followed by condensation as a separation process. Also, underwater mass diffusion to the steam bubbles should work as an important part of degassing, especially to reach a lower level of DO concentration.

1.6 COMPARISON OF CONVENTIONAL HIGH-TEMPERATURE-DRIVEN DEAERATION VS. LOW-PRESSURE-DRIVEN DEAERATION

In order to understand the basic deaeration mechanism, let us consider a case of spray-type pressure deaerator, in which preheated water is sprayed through a nozzle to be heated and evaporated in a steam-filled space. Other than being heated and evaporated, mass diffusion between the droplet and the steam space may be conceivable, as illustrated in Figure 1.3(a), where dissolved gas (mostly oxygen and nitrogen) in the droplet may diffuse to the steam space, where the gas concentration is very low (this is to be shown in the following chapter). Also, after the vapor is condensed and contained in the vessel, steam that is generated through a separate boiler is supplied to the steam sparger in order for steam bubbles to collect any remained oxygen contents in the bulk water, as shown in Figure 1.3(b). A conventional deaerator requires a water heater, a pressurization system for spray nozzles, a separate boiler system for steam generation, and a main water tank. The resulting system is expected to be bulky, be complex in structure, and consume significant amounts of energy.

Now, if we look at degassing options under low pressure, two interesting physical phenomena come to mind. The first is its behavior according to Henry's law, which states that the solubility of a gas is proportional to the partial pressure of that gas. Lowering the pressure surrounding the water reduces the solubility limit of the gas, which can cause the gas in the water to become supersaturated. In this case, bubbles may appear below the liquid surface level without any special measures being taken. The essence of those bubbles is mainly a mixture of supersaturated gaseous solutes, to which vapor corresponding to the saturated vapor pressure at a given temperature will be added. This phenomenon can be likened to wringing out a wet towel and the remaining moisture that the towel cannot contain falls out (Figure 1.4). These supersaturated solutes separate relatively easily in the form of bubbles through any kind of disturbance, such as mixing or impact.

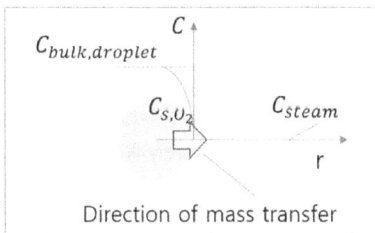

(a) Droplet diffusion

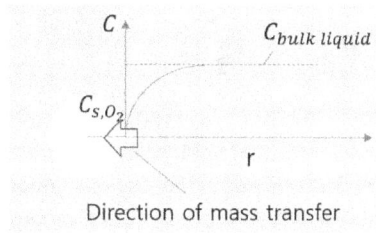

(b) Steam bubble diffusion

FIGURE 1.3 Diffusion models in conventional thermal deaeration process; (a) droplet diffusion and (b) steam bubble diffusion.

FIGURE 1.4   An image of wringing water out of a towel.

It should be noted that degassing based on solubility limit naturally would release gases based on the fraction of individual gases dissolved upon the solubility of each gas component. To simplify the discussion, let us assume that the atmosphere is composed of 79.1% nitrogen and 20.9% oxygen. Noting that the Henry's solubility constant for oxygen (1.28×10^{-3} mol/L·atm at 25°C) is about double that for nitrogen (6.48×10^{-4} mol/L·atm at 25°C) [17], the amount of dissolved oxygen and nitrogen under a saturated condition can be roughly estimated to be $1.28 \times 10^{-3} \times 0.209 = 2.7 \times 10^{-4}$ mol/L and $6.48 \times 10^{-4} \times 0.791 = 5.13 \times 10^{-4}$ mol/L, respectively, which results in the oxygen volume fraction of dissolved gas under 1 atm of idealized dry air to be $2.7 / (2.7 + 5.13) = 34.5\%$. Actual experimental measurements had been conducted and reported with supporting results [7]. Note that in this discussion, the contribution of vapor is not considered just for simplicity, but the discussion considering the vapor contribution is continued in detail in Sec. 2.1.3.

The second is the creation of vapor bubbles due to phase change and the resulting mass diffusion arising from the low dissolved gas concentration inside the vapor bubbles. This is a key part of all the stories presented in this book. It might be my lack of understanding, but it seems that not much academic research has been done so far on the creation and behavior of room-temperature vapor bubbles, and there did not appear to be many recognized research results. My main interest, and the difference from the research results of existing researchers, is the generation of vapor bubbles at low pressure and the properties of the vapor bubbles created. In particular, academic discussion of achievable levels of degassing in relation to mass transfer is not new but is not commonly addressed in published literature. In the power deaerator system, the so-called deaeration section is the part where deaeration occurs by mass transfer between the steam bubbles supplied through the steam sparger and the bulk water. Although a clear explanation of this part was not documented in most commercial open literature, the degassing principle is understood here as taking advantage of the very low oxygen concentration inside the steam bubbles to discharge the dissolved gas in the bulk water by concentration difference (Figure 1.3(b)). This degassing method using high-temperature steam bubbles can be applied to vapor bubbles created in a vacuum, with the expectation of equal or better performance. The second mechanism by which bubbles are created in a vacuum happens when a liquid undergoes a phase change near its saturated vapor pressure. For example, water has a saturated vapor pressure of 3.17 kPa at 25°C, below which the liquid phase is no longer stable and is ready to evaporate. Bubbles created by lowered pressure are generally accepted as cavitation bubbles and are known to have negative effects, such as noise, vibration, and material removal due to implosion when they appear and disappear [18]. However, this negative perception is only valid when there is a large pressure difference between the vaporization point and its neighbors, for example, the back of the propeller (suction side) and the nearby bulk water, as shown in Figure 1.5(a). Here, the pressure difference between a cavitation bubble and its neighbors can be more than 100 kPa. When the pressure near the vaporization site is low enough, close to the saturated vapor pressure (Figure 1.5(b)), the creation and collapse of vapor bubbles may not perform hazardous or harmful actions. When a vapor bubble is created, the amount of dissolved gas (in moles or mass) remains the same as originally dissolved in the liquid, but its volume expands by a factor of 10^4. Due to this volume expansion of water, the volume fraction of originally dissolved gases such as oxygen and nitrogen is greatly reduced. Additionally, because the pressure level during bubble generation is much lower than atmospheric pressure, the partial pressures of dissolved gases in the vapor bubble are approximately two orders of magnitude lower than that of hot steam. (Calculation procedures for specific concentrations are introduced in Take a Break! 2).

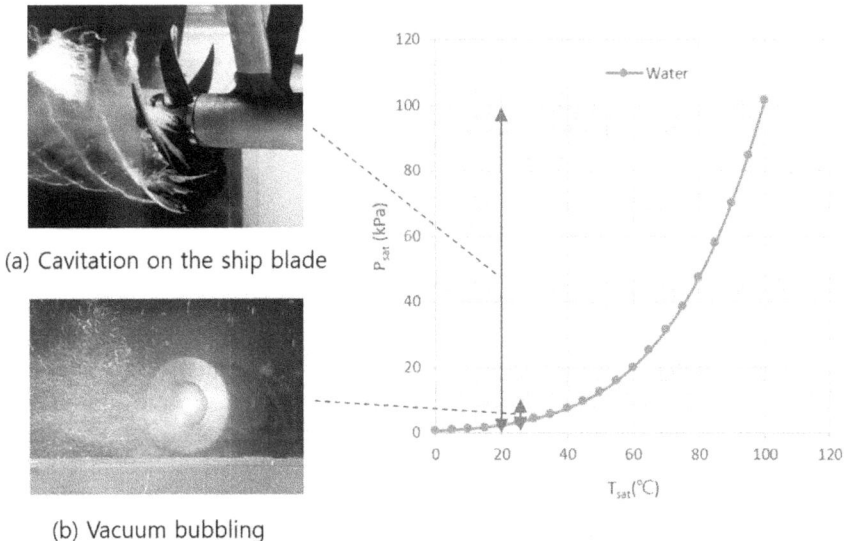

(a) Cavitation on the ship blade

(b) Vacuum bubbling

FIGURE 1.5 Comparison of two different cases of cavitation.

TAKE A BREAK! 1

If there is 3.13% water vapor in the air at 25°C, how many moles of oxygen, nitrogen, and water vapor are there in 1 m³ of atmospheric air? Assume that dry air is composed of 20.9% oxygen and 79.1% nitrogen.

TALETELLER 1

Since there is 3.13% water vapor in the atmosphere and the idealized dry air consists of 20.9% oxygen and 79.1% nitrogen, the volume fractions of each gas are:

$$Y_{vap} = 0.0313$$

$$Y_{O_2} = (1-0.0313) \times 0.209 = 0.2025$$

$$Y_{N_2} = (1-0.0313) \times 0.791 = 0.7662$$

Now, the number of moles of each component, N_i, can be obtained from the equation of state:

$$N_i = \frac{Y_i pV}{R_u T}$$

where the universal gas constant $R_u = 8.314$ J/mol·K, $p = 101.3$ kPa, $V = 1$ m³, and $T = 273.15 + 25 = 298.15$ K. When applied to water vapor, oxygen, and nitrogen, respectively,

we get $N_{vap} = 1.28\,mol$, $N_{O_2} = 8.28\,mol$, and $N_{N_2} = 31.3\,mol$, respectively.

The total number of moles is 40.86 mol, which is slightly smaller than the number of moles that is expected in the same volume at 0°C and 1 atm, 1 m³/ (0.0224 m³/mol) = 44.64 mol.

1.7   ADVANTAGES OF USING VAPOR BUBBLING

The two principles applicable to degassing from liquids, solubility and phase change, both depend on temperature and pressure fields, and so far the upper right region (high-temperature and high-pressure region) of the temperature–pressure diagram (Figure 1.2) has been extensively used, while not much attention has been paid to the degassing process in the low-temperature, low-pressure region at the bottom left of the figure, which can save energy with a simple structure. In the case of high-temperature systems, a significant amount of energy input is required because the working fluid must be directly heated even before vaporization or bubble generation, but the vacuum bubbling method generates bubbles through vacuum without a separate heating process, resulting in a dramatic reduction in energy costs. In the case of vacuum bubbling, no additional energy is required for the system except for the pdV work required to make empty space a vacuum, and the vacuum level required for the process is in the range of 1 to several kPa, which is not difficult at all, considering the current level of technology. By applying vacuum bubbling to the degassing process, degassing can be accomplished with a simple configuration, and the operating principle is clear. As will be explained later, the oxygen concentration inside the vapor bubbles generated at low pressure is close to 0, so high-performance degassing by bubble diffusion seems feasible. This is because the higher specific volume of the vapor bubbles produced at very low pressures results in much more favorable mass transfer conditions (lower dissolved gas partial pressures in the bubbles) than at higher pressures. Deaeration by the vapor bubbles generated in this way appears to have no problem achieving the highest known industrial deaeration level of 5 ppb. Additionally, in the process of implementing this process, the main control variable is the pressure inside the tank, so it has the advantage of being easy to manipulate or automate. The most important advantage of the vacuum bubbling process is that the expected energy requirement is significantly lower than that of existing methods, but in addition, it is simple to construct, has low initial cost, is easy to operate, and does not require third-party material intervention. Therefore, it can be said that this approach needs to be considered as an economical, environmentally friendly, and sustainable alternative to the highest level of performance, at least in the field of degassing. So why not consider this process, which can provide equal or better performance? My goal in writing this manuscript is to guide you through the theoretical background related to the vacuum bubbling process and introduce its important current and future application possibilities.

1.8   COMPARISON AGAINST THERMAL DEAERATION PROCESS

The phase change from liquid to gas is expressed as evaporation or boiling. In heat transfer, the process of increasing temperature when heat energy is applied is called heating, and applying heat to a temperature higher than the boiling temperature is called superheating. However, if we look at the heating process from a more general perspective, we see that other phenomena than heat transfer are also involved. One of them is that as the temperature of a liquid increases due to heating, the solubility decreases (Figure 1.1). In Figure 1.1, which shows the change in solubility of oxygen in water as a function of temperature and pressure, we see that when the temperature is raised to 100°C (212°F) at atmospheric pressure, the solubility changes to almost zero. (Of course, the temperature at which solubility reaches zero will change as the pressure inside the container is higher or lower than atmospheric pressure.) Therefore, in the thermal degassing process, the liquid is heated to reduce the amount of gas it can hold, that is, its solubility. At this time, spraying or trays are used as a method to provide the maximum surface area so that the supersaturated dissolved gas can escape from the liquid more easily. The next step is to proceed with additional degassing through mass diffusion by sparging externally generated steam in water as a method to remove dissolved gases that have not yet been deaerated. At this time, although there is no separate explanation, the oxygen concentration of the steam is assumed to be very low.

Vacuum bubbling is very similar in principle to thermal degassing. Just as the thermal degassing process raises the temperature to create a low solubility state, vacuum bubbling lowers the pressure to create a low solubility state based on Henry's solubility law. In addition, by generating vapor bubbles in water under lower pressure, the sparging effect using steam in the case of thermal degassing process is achieved through self-generated vapor bubbles. In this way, it can be said that the operating mechanism of heated degassing and vacuum bubbling is almost the same, but if you look more closely, you can learn about various unique points of vacuum bubbling. For example, in the process of reducing solubility, vacuum bubbling simply depressurizes the empty space within the container rather than adding heat to the water itself. In addition, diffusion mass transfer through vapor bubbles does not rely on steam supplied from an external boiler but uses vapor bubbles generated inside the fluid by a bubbler. By doing this, the necessary equipment and accompanying energy demand can be significantly reduced. Although thermal degassing and vacuum bubbling can be said to share almost the same operating mechanism, several unique features can be found, as shown in Table 1.1.

As mentioned earlier, in the process of reducing solubility, vacuum bubbling only depressurizes the empty space in the container without heating the water itself. This can be expected to have the effect of dramatically reducing the required equipment and resulting energy demand. For a better understanding, it would be interesting to compare the energy required for degassing water by thermal degassing versus degassing it by vacuum bubbling. For the sake of this discussion, let us consider the following simple comparison example: Suppose that 400 L of water at room temperature of 25°C is to be degassed. In the case of the thermal degassing process, all 400 L of water should be heated, evaporated, and then condensed again. Then, deaeration with additional steam proceeds. To estimate the energy required for the degassing process, one can first calculate the amount of energy consumed to heat and evaporate the water. If heating and evaporation occur under an atmospheric pressure, the total

TABLE 1.1

Comparison of Thermal vs. Vacuum Bubbling Deaeration Process

| | Implemented Methodology (Physical Device*) | | | |
	Solubility-Driven Deaeration	Diffusion-Driven Deaeration	Energy Consumption For	Initial Cost For
Thermal process	Heating (pre-heater)	Steam sparging (external boiler and steam sparger)	Heating, vaporizing, pressurizing, steam generation	Vessel, heat exchanger, spray system, tray, etc.
Vacuum bubbling process	Vacuuming (vacuum pump)	Vapor bubbling (bubbler)	Vacuuming, water pumping	Vessel, vacuum pump, pump-driven bubbler

Note: *Not including monitoring sensors and controllers.

energy required is the sum of the energy required for heating and for evaporation. That is:

$$E = Q_{heating} + Q_{evap} = m\left(c\Delta T + h_{fg}\right)$$

Now, assuming the density of water at 25°C is 997 kg/m³, the specific heat is $c = 4.18$ kJ/kg·K, $h_{fg} = 2256.5$ kJ/kg [19], then, the total required energy is:

$$E\left(kWh\right) = 997 kg / m^3 \times 0.4m^3 \times \left(4.18kJ / kg \cdot K \times 75K + 2256.5kJ / kg\right)$$
$$= 1025035 kJ / \left(3600 kJ / kWh\right) = 284.7 kWh$$

Additionally, data on the amount of steam to be supplied from outside is required. For this discussion, we refer to the vent rate in the random sample of the Steam System Modeling Tool (SSMT) provided by the Department of Energy Efficiency and Renewable Energy [20]. Since the range of 0.3% to 1.0% of feedwater is presented, assuming about 0.5% of this, and assuming that the state of the steam exiting the vent is saturated at 100°C, the energy consumed by the vent is:

$$E_{vent}\left(kWh\right) = 997 kg / m^3 \times 0.4m^3 \times 0.005 \times 2256.5kJ / kg = 1.25 kWh$$

In other words, the total energy consumed to heat, evaporate, and degas 400 L of water at 25°C is 286.0 kWh or 715 kWh/m³. Let us now consider the case of degassing by reduced pressure. Assuming that the initial ullage volume and initial pressure of the tank are 21 L and 1 atm (= 100 Pa), respectively, and considering the case of decompressing to 0.01 atm (= 1 Pa) under isothermal conditions, the energy required for decompression is:

$$W = \int_{V_1}^{V_2} pdV = p_1V_1\int_{V_1}^{V_2} \frac{dV}{V} = p_1V_1 ln\left(\frac{V_2}{V_1}\right)$$

In this case, if $p_1 = 100$ kPa, $V_1 = 0.021$ m³, and $p_2 = 1$ kPa, $V_2 = 2.1$ m³ are substituted in the equation, the work required for decompression is:

$$W_{vac} = 100 kPa \times 0.021 m^3 \times ln\left(\frac{2.1}{0.021}\right)$$
$$= 9.67kJ / \left(3600kJ / kWh\right) = 0.00269 kWh$$

For reference, as can be seen from the calculation process, the energy required for decompression is proportional to the initial volume. After pressure reduction, a bubbler driven by a water pump is required. For example, let us say ten hours of degassing is required with a power of 20W (this is the power used for degassing in the author's laboratory). The power consumed for pumping is:

$$W_{pump} = 20W \times 10h = 0.2 kWh$$

As a practical comparison for this discussion, an experiment equivalent to the preceding example was performed, resulting in a total energy usage of 1.67 kWh [16]. The main reason for the difference between the initial prediction and the experimental results was the additional operation of vacuum pump due to the continued generation of bubbles. Comparing the total energy required for degassing in this sample case shows two orders of magnitude less energy usage for vacuum bubble degassing compared to a simplified thermal degassing model represented by heating, evaporation, and condensation, followed by steam sparging or steam stripping.

1.9   THE WHOLE PICTURE OF VACUUM BUBBLING

A closer look at the process of temperature rise and phase change when heat is applied to a system will help in understanding the expected phenomena caused by depressurization. In the initial stage of heating, dissolved gases are separated from the liquid due to a decrease in solubility with increasing temperature. In the process of thermal degassing, a means for promoting heat transfer or mass transfer, such as generating liquid droplets or creating a flow through a tray, is used to help the supersaturated dissolved gas separate from the liquid. When the external heating temperature is higher than the saturation temperature and becomes a superheated state, vapor bubbles are generated from the hot surface. At this time, according to the temperature conditions of the surrounding liquid, when the temperature of the bulk liquid has not yet reached the saturation temperature, the generated vapor bubbles immediately shrink or disappear upon departure from the heated surface, and when the ambient temperature reaches the saturation temperature, the generated vapor bubbles may not shrink and rise due to buoyancy from the liquid, as illustrated by nucleate boiling region in Figure 1.6. The volume fraction of the non-condensable gas inside the vapor bubble is very low due to the volume expansion of the liquid, and mass diffusion occurs due to the concentration difference between the vapor or steam bubble and the bulk body of liquid. If a large amount of thermal energy is continuously supplied to the system, the phase change continues in the form of film boiling.

In the case of vacuum bubbling, the dissolved gas is separated from the liquid due to the decrease in saturated solubility as the internal pressure of the vessel decreases. However, there is a limiting pressure condition in which Henry's solubility law may no longer be applicable. When vapor bubbles are generated as a consequence of phase change at pressure of saturated vapor pressure or less, the specific volume of the vapor dramatically increases. Mass diffusion with the surrounding liquid driven by the concentration difference then occurs due to the very low concentration of non-condensable gas inside the vapor bubble, as in the case of steam by heating. Finally, if the temperature and pressure conditions of the liquid lie in the so-called metastable region, a necessary condition for the generation of vapor bubbles, it will be possible to generate large amounts of vapor bubbles in proportion to the amount of energy applied to the system according to the principle of energy conservation. In this context, vacuum bubbling can be thought of as being divided into the following three stages.

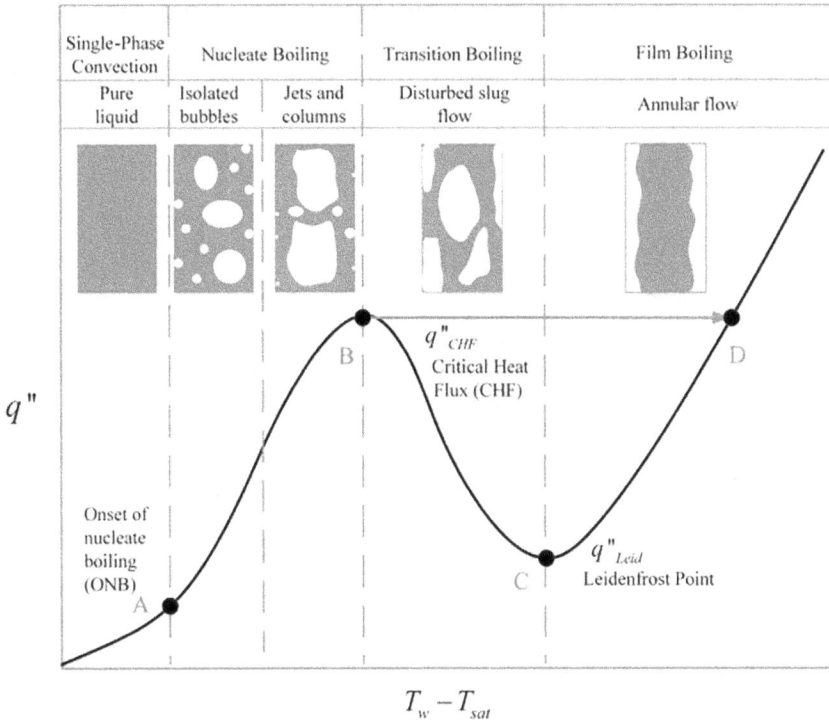

FIGURE 1.6 A typical boiling curve with respect to the degree of superheating [21].

1.9.1 BUBBLING DRIVEN BY SOLUBILITY LIMIT (PHASE 1)

Although degassing occurs through bubbling, the manner in which deaeration is performed can be quite different in nature. At a given temperature, when the pressure level in the liquid is greater than the saturated vapor pressure, voiding is the result of a process in which a supersaturated solute seeks equilibrium at a lowered pressure. The contents of the bubble are supersaturated solute and saturated vapor, and we can prove this by measuring the oxygen concentration of the extracted gas or air. The air extracted from water showed oxygen concentration of 30% or higher, which is close to the ideally expected oxygen concentration of 34.5% [7].

1.9.2 BUBBLING OF VAPOR (PHASE 2)

If the pressure anywhere in the liquid is lowered to the vapor pressure or lower, vapor bubbles may be generated in addition to bubbles due to solubility. In phase 1 bubbling, supersaturated solute escapes as bubbles due to the limit of solubility, but in phase 2 bubbling, liquid is vaporized locally and bubbles are formed. In this case, it is a fundamentally different phenomenon from the phase 1 bubbling in that mass diffusion occurs due to the difference between the concentration of gas inside the vapor bubble and that of dissolved gas in the liquid. For example, when water is vaporized into bubbles, the oxygen concentration inside the bubbles becomes about 10^{-6} in terms of volume fraction (See Take a Break! 2), and oxygen dissolved in the liquid diffuses toward the bubbles and degassing occurs. Therefore, in this case, the generation of bubbles is most important, but the bubble behavior (growth, contraction, loss), size, residence time, etc. can be important variables in the consideration of degassing.

1.9.3   Massive Vapor Bubbling (Phase 3)

Although there are not many open literature discussing the generation and maintenance of room-temperature vapor bubbles [22, 23], the author analyzed the requirements for bubble generation and confirmed that a large amount of bubbles were generated through experiments while writing this article. According to the research results so far [16], the necessary condition for vapor bubbles to be created and maintained at room temperature requires the pressure at the location (depth) where the bubbles are to be generated must exist at least in the metastable region. (An explanation of the metastable region is presented in Sec. 2.2.1). When a liquid exists in the metastable region under reduced pressure, a phase change from liquid to vapor can already occur in thermodynamic state-wise, represented by temperature and pressure. However, it can be seen that additional energy supply for phase change is required. Thus, if the liquid is in the metastable region, it will be possible to ensure that a significant portion of the additional energy supply is used for the phase change. Ideally, therefore, it can be assumed that all energy supplied from outside the system is used as heat of vaporization. And a model, in which the energy required for vaporization is supplied in the form of work rather than heat energy, could be a model of vacuum bubbling. According to this model, a massive vapor bubbling may result through non-thermal mechanical work in a liquid in the metastable region.

To summarize, as shown in Figure 1.7, vacuum bubbling can be largely divided into three phases. The first is the phase in which the supersaturated solutes escape in the form of bubbles due to the decrease in saturation solubility with reduced pressure (Figure 1.7(a)). In this case, the pressure within the fluid ranges between the atmospheric pressure and the saturated vapor pressure. Degassing applications that can exploit this phenomenon include extracting high-oxygen-concentration air dissolved in water for respiration or storage. (Refer to artificial gills in Sec. 3.3). It can also be applied for pre-deaeration of jet fuel to prepare for unexpected explosion safety or to control the oxidation of biodiesel. The level of deaeration required for jet fuel is not so high, so it seems that first-phase bubbling seems sufficient, but in the case of deaeration to prevent oxidative deterioration of biodiesel jet fuel, which has recently attracted a lot of attention, the next phase of bubbling may be required. The second phase of vacuum bubbling is a phase in which vapor bubbles are created and the dissolved gas in the liquid diffuses into the bubbles (Figure 1.7(b)). A prerequisite for this phase is that the pressure at the location of the bubble generation must be in the metastable region below the saturated vapor pressure. Since the volume fraction of the non-condensable gas inside the vapor bubble is almost zero, it is a very favorable condition for mass diffusion, and theoretically, the concentration of dissolved gas in the liquid can be realized down to zero. Therefore, it will be

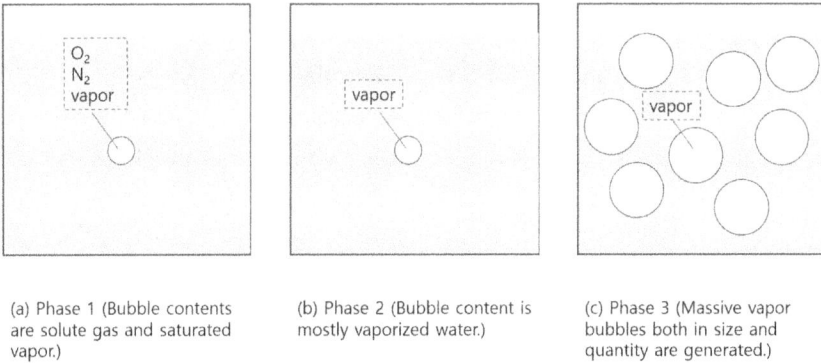

(a) Phase 1 (Bubble contents are solute gas and saturated vapor.)

(b) Phase 2 (Bubble content is mostly vaporized water.)

(c) Phase 3 (Massive vapor bubbles both in size and quantity are generated.)

FIGURE 1.7   Three phases of vacuum bubbling; (a) Phase 1 (Bubble contents are solute gas and saturated vapor), (b) Phase 2 (Bubble contents are mostly vaporized water), and (c) Phase 3 (Massive vapor bubbles both in size and quantity are generated).

an effective and economical alternative that can satisfy the requirements of most industrial and research applications. However, since the basic mechanism of the second-phase degassing is mass diffusion, the rate of deaeration is noticeably slower than that of the first phase. In addition, it is more closely related to the problems of bubble dynamics, such as the amount, size, and location (depth) of bubbles. The third phase of vacuum bubbling is the phase in which vapor bubbles are generated on a large scale (Figure 1.7(c)). Under the condition of the second phase, vapor bubbles are generated in proportion to the supplied energy, so if a large amount of energy is supplied, a larger amount of vapor bubbles may be created, at least theoretically. This is a concept similar to vapor generation by film boiling in the boiling model through heating, as shown in Figure 1.6. That is, sufficient mechanical work is supposed to be supplied to the liquid system to generate large amounts of vapor. At this phase, mass evaporation is more meaningful than bubble formation, and it will be applicable to low-temperature desalination through low-temperature evaporation. The amount of energy required for vapor generation is not significantly different from that of the thermal desalination process; however, the important difference is that in vacuum bubbling, instead of burning fossil fuels for high-temperature heating, low-temperature heat energy or an equivalent amount of electrical energy can be used instead.

1.10   BUBBLING IMPLEMENTATION AND MODEL

So far, we have learned about the concept and overall characteristics of vacuum bubbling. Before implementing this in practice, let us take a look at the availability of vacuum technology. In the case of vacuum bubbling introduced in this book, the main liquid to be applied is water, and the vapor pressure of water at room temperature ranges from 1 kPa to several kPa. And the conditions for creating vapor bubbles, which will be introduced in Sec. 2.2, correspond to a metastable region at or below saturated vapor pressure, so a vacuum of approximately 1 kPa is required. Advances in vacuum technology seem to have grown rapidly in recent industrial

TABLE 1.2
Degrees of Vacuum and Their Pressure Boundaries [24]

	Pressure Boundaries (mbar)	Pressure Boundaries (Pa)
Low vacuum (LV)	$1000 \sim 1$	$10^5 \sim 10^2$
Medium vacuum (MV)	$1 \sim 10^{-3}$	$10^2 \sim 10^{-1}$
High vacuum (HV)	$10^{-3} \sim 10^{-9}$	$10^{-1} \sim 10^{-7}$
Ultra-high vacuum (UHV)	$10^{-9} \sim 10^{-12}$	$10^{-7} \sim 10^{-10}$
Extreme vacuum (EV)	$< 10^{-12}$	$< 10^{-10}$

developments, especially in connection with the development of high-tech products such as semiconductors. As a result, the 1 kPa level of vacuum required for vacuum bubbling is classified as low vacuum [24], and this level of vacuum can be achieved with a routine diaphragm-type vacuum pump. Table 1.2 presents the classification and pressure range of vacuum as seen in the vacuum field. Therefore, concerns about the vacuum environment, which was considered one of the technical challenges for industrial application of vacuum bubbling technology, are no longer a major headache.

How to implement vacuum bubbling at room temperature, and especially the process of applying it to design, can be accomplished in various forms, so here we will introduce the basic configuration and principles. Among various configuration options, we will introduce a relatively simple method of configuring a vacuum bubbling system through Figure 1.8, one of the examples. This model was proposed for a dissolved oxygen removal device in a power plant cooling system.

1.11   IMPLEMENTATION

One way to effectively apply the two principles mentioned earlier is to generate bubbles using a separate pressure-reducing device while sufficiently lowering the pressure inside the container. The reason for sufficiently lowering the pressure in the container is to minimize the solubility of dissolved gases according to Henry's law and maximize the discharge of supersaturated gases. This is also to minimize the power required for the next step, vapor bubble generation. Figure 1.8, which follows, shows the basic structure of a vacuum bubbling deaerator.

Inside the pressure-resistant sealed container, which can withstand decompression of down to about 1 kPa, there is a bubbler assembly consisting of a submersible pump and a venturi nozzle, and a ventilation port for decompression is connected to the vacuum pump at the top of the container. Since the operation of a vacuum pump must be regulated according to the pressure level inside the vessel, it is useful to have an automatic valve opening and closing function linked to a pressure sensor. A feedwater inlet is installed at the top of the container, and a treated water outlet is installed at the bottom. Additionally, a power cable to drive the submersible pump and sensors to measure temperature, pressure, and dissolved oxygen concentration are added. It is advisable to install a visualization window to monitor the situation inside the vessel.

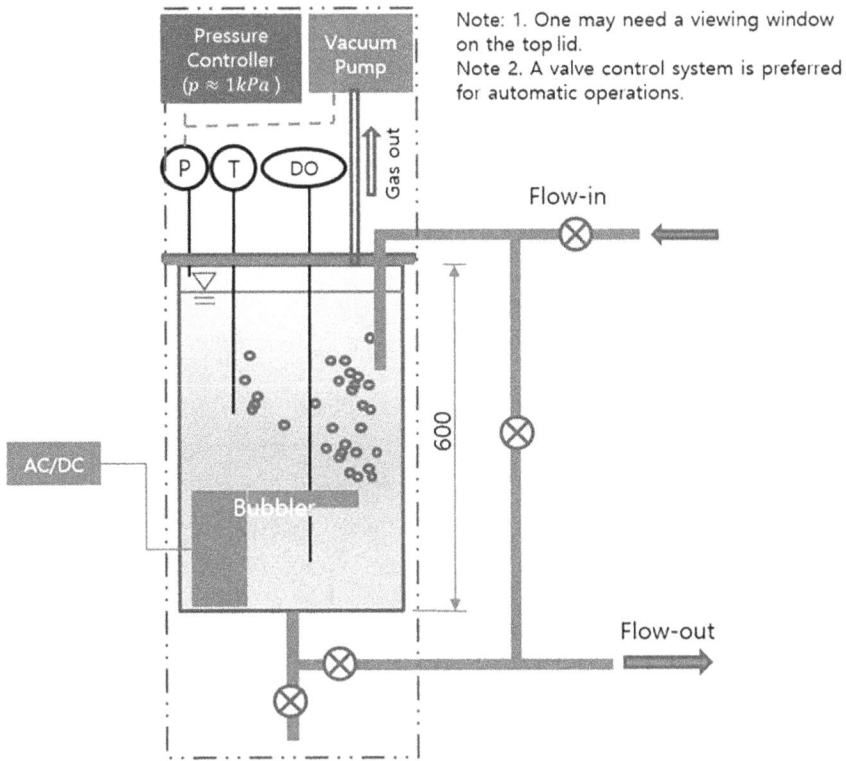

FIGURE 1.8   Illustrative image of a proposed vacuum bubbling deaerator.

The operation method is to first fill the container with water to be degassed through the inlet pipe, then close all other connected valves and maintain the required level of vacuum using a vacuum pump. At the same time, or at different times, the bubbler driven by a submersible pump is operated. As the bubbler operates, bubbles are generated in the water, and when these gases rise, the pressure inside the container increases. When the pressure in the container rises above the preset value, the conditions for generating bubbles change, so the vacuum pump is operated until the pressure inside the container reaches the lower limit of the preset pressure range. This task can be automated by adding an automatic control function.

1.12   HOW IT WORKS

Now, we will introduce the operating principle of this degassing device based on the phase-change diagram of water shown in Figure 1.9. In this approach, decompression occurs in two stages. The first stage of decompression is to lower the pressure in the headspace of the vessel. For example, the pressure of the vessel can be lowered to 3.17 kPa, which is the saturated vapor pressure at 25°C. (Usually, the lower limit of pressure depends on the power of the vacuum pump.) This lowers the pressure in the space surrounding the water, and the solubility of the gas according to Henry's law

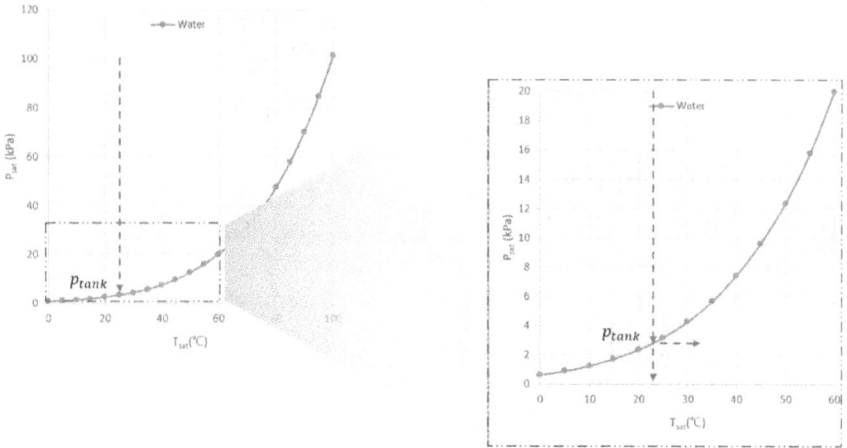

FIGURE 1.9 Two-step approaches for vacuum bubbling.

also decreases in proportion to the partial pressure of each gas. Then, some of the dissolved gases that were initially dissolved at atmospheric pressure become supersaturated solutes because the amount of solutes that can be dissolved decreases due to reduced pressure, and these solutes are no longer stably dissolved in the solution and bubbles may form in the solution. For example, as shown in Figure 1.10, when a solution that was saturated under atmospheric pressure is reduced to 0.1 atm, theoretically, 90% of the originally dissolved solute is destined to be discharged in the form of bubbles. As mentioned earlier, the creation of bubbles can be seen as occurring in the process of a supersaturated solute finding its equilibrium state, and the generation of bubbles can be promoted by providing a cause for appropriate mixing. The inside of the bubbles created in this way is composed of saturated vapor as well as dissolved oxygen and nitrogen. The advantage of this method of creating a vacuum inside the vessel is related to the energy required to reach saturation, which is the phase-change boundary. In the case of the heating method, the entire water inside the vessel must be heated, while in the case of decompression, the pressure in the empty space within the vessel is reduced. In other words, it does not change the state of the water itself but only changes the surrounding conditions (pressure).

Even if the pressure reaches the saturated vapor pressure, water does not vaporize immediately. There are two extreme ways to create vapor in this state: local heating and local decompression. Local heating is a method of causing a phase change through local superheating (heating in the direction of the arrow on the right in the saturated state in Figure 1.9). The advantage of this method is that a phase change can be achieved through low-temperature heating. Meanwhile, local decompression is to create a tension state by making the pressure lower than the saturated vapor pressure to create cavitation (decompression in the direction of the downward arrow in the saturated state in Figure 1.9), and in this book, the latter method is to be discussed. It is worth reminding that there are also approaches that use surface modification to create bubbles more easily. One way to achieve localized pressure reduction is by

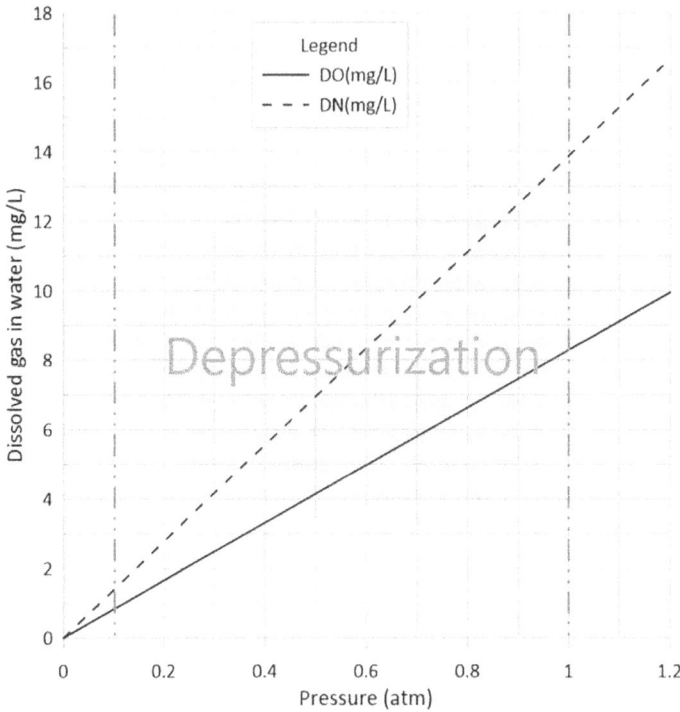

FIGURE 1.10   Expected dissolved gas contents (mg/L) of oxygen and nitrogen in water at 25 according to Henry's law under an equilibrium state.

having a flow through the venturi nozzle by means of a submersible pump installed underwater, and the actual bubbles are generated in the nozzle throat of the venturi where the pressure is the lowest. At this time, the lower the tank pressure, the less the power required to reduce the additional pressure required to generate bubbles. This is the energy consumed directly in generating bubbles, so it is important to find optimal conditions. Details related to this will be covered in the section on vapor bubble generation in Chapter 2.

I would like to raise a point regarding the application limit of Henry's solubility law. It involves the pressure being lowered sufficiently so that a phase change occurs at or below the saturated vapor pressure of the solution. Below the phase-change pressure mentioned in relation to Figure 1.9, the liquid can change phase into vapor depending on the amount of energy supplied, so the solubility behavior according to pressure change reaches the applicable limit. In other words, when the local pressure becomes lower than the saturated vapor pressure and vaporization occurs, the law of solubility for the solution can no longer be applied. Considering the limitations of Henry's law, the water solubility behavior shown in Figure 1.10 should be newly considered, as shown in Figure 1.11. Of course, just because a solution reaches the saturated vapor pressure or falls below it does not necessarily mean that it vaporizes

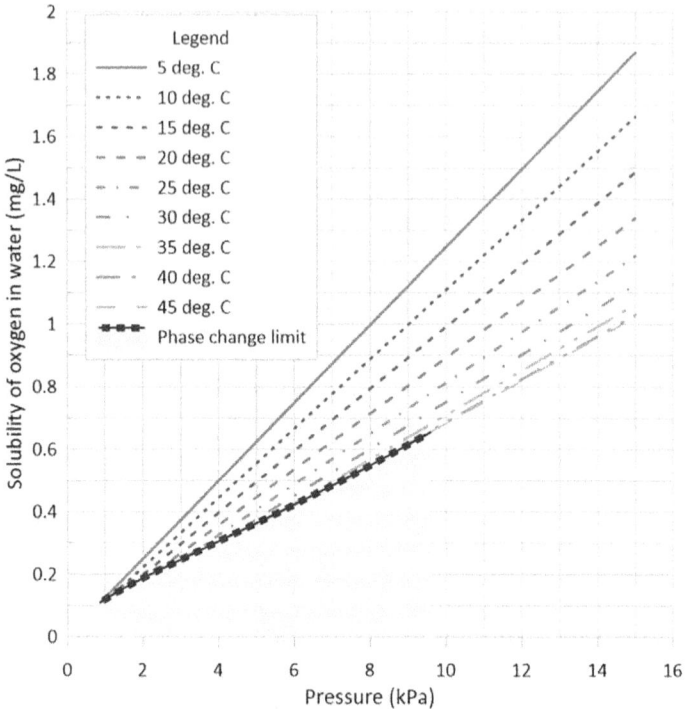

FIGURE 1.11   Solubility of oxygen in water with phase change limits.

(explained in detail in Sec. 2.2.1), but once the liquid phase changes to vapor, separate consideration is needed.

As shown in Figure 1.9, the second step of vacuum bubbling is to create vapor bubbles through additional local decompression in the reduced pressure state of step 1. Unlike the previously mentioned supersaturated solute that separates from the liquid in the form of bubbles, when a part of the liquid reaches a pressure below the saturated vapor pressure through additional decompression, the liquid may undergo a phase change and vaporize. At this time, the pressure difference between the lowered pressure and the saturated vapor pressure is called tension, and evaporation is known to depend on the magnitude and duration of this tension [23]. This is the same situation as water existing in both liquid and vapor phases even under the same thermodynamic state at 1 atm pressure and 100°C, and water evaporates depending on the amount of energy required for evaporation. When water evaporates and becomes steam, what phenomenon can we expect? This is not only a very important issue academically but also in terms of application, but unfortunately, it has been difficult to find in the published literature on the application limits of Henry's law so far. As introduced in "Take a Break! 2," if water that was in equilibrium with the atmosphere at room temperature 25°C and 1 atm was heated and evaporated (to simplify the problem, we boldly assume that there is no dissolved gas escaping through bubbles during heating), the oxygen volume fraction of this steam can be estimated to be 4.69×10^{-6}. In reality, this figure may be lower because most of the dissolved gases will have already escaped due to a decrease in solubility during heating, and furthermore, if evaporation takes place at a pressure as low as 1 kPa, we can expect an even lower level of oxygen in terms of its partial pressure. So how much lower can it go? The author believes that one of the most important aspects of writing this book was overcoming misunderstandings related to vapor bubbles. In other words, vapor bubbles have an oxygen concentration close to zero and can have a gas stripping effect comparable to or better than the purity of the inert gas generally used for gas stripping.

TAKE A BREAK! 2

Suppose we have 1 L of water at 25°C that is in equilibrium with the atmosphere. If this water is heated and vaporized under 1 atm (101.3 kPa), what would be the oxygen concentration when the water is completely evaporated? Neglect degassing during the heating up to vaporization.

TALETELLER 2

Let us assume Henry's solubility constant of 1.28×10^{-3} mole/(L·atm) for oxygen and 6.48×10^{-4} mole/(L·atm), respectively, for water at 25°C and the partial pressure of saturated water is 0.0313 atm or 3.17 kPa. The water is assumed to be in equilibrium with an idealized atmosphere of 20.9 % oxygen and 79.1 % nitrogen. Note that the gas constants for oxygen, nitrogen, and vapor are given to be 0.2598 kJ/kg, 0.2968 kJ/kg, and 0.4615 kJ/kg, respectively.

- Initial water volume: $1 L = 0.001 \ m^3$
- Concentration of dissolved oxygen in water at 25°C:

$$G_{O_2} = H_{O_2} p_{O_2} = 1.28 \times 10^{-3} mole / L \cdot atm \times 1 atm \times (1 - 0.0313) \times 0.209$$
$$= 0.00026 \, mole / L \times 32 g / mole \times 1000 mg / g = 8.29 mg / L$$

- Concentration of dissolved nitrogen in water at 25°C:

$$G_{N_2} = H_{N_2} p_{N2} = 6.48 \times 10^{-4} mole / L \cdot atm \times 1 atm \times (1 - 0.0313) \times 0.791$$
$$= 0.00050 \, mole / L \times 28 g / mole \times 1000 mg / g = 13.90 mg / L$$

1. Once this water is completely evaporated through heating and is in the state of 100°C and 1 atm, then:
   - Volume occupied by the oxygen:

$$V_{O_2} = \frac{m_{O_2} R_{O_2} T}{p} = \frac{8.29 \times 10^{-6} kg \times 0.2598 \, kJ \, / \, kg \cdot K \times 373.15 K}{101.3 \, kPa}$$

$$= 7.93 \times 10^{-6} m^3 = 0.00793 \, L$$

- Volume occupied by nitrogen:

$$V_{N_2} = \frac{m_{N_2} R_{N_2} T}{p} = \frac{13.9 \times 10^{-6} kg \times 0.2968 \, kJ \, / \, kg \cdot K \times 373.15 K}{101.3 \, kPa}$$

$$= 15.20 \times 10^{-6} m^3 = 0.01520 \, L$$

- Vapor volume:

$$V_{vap} = \frac{m_v R_v T}{p} = \frac{0.997 \, kg \times 0.4615 \, kJ \, / \, kg \cdot K \times 373.15 K}{101.3 \, kPa}$$

$$= 1.695 \, m^3 = 1,695 \, L$$

- Oxygen concentration of vapor mixture:

$$C = \frac{V_{O_2}}{V_{mix}} = \frac{V_{O_2}}{V_{O_2} + V_{N_2} + V_{vap}} = \frac{0.00793}{0.00793 + 0.01520 + 1695} = 4.68 \times 10^{-4} \%$$

2. If the water is evaporated through vacuuming, for example, using vacuum bubbling, and is in the state of 25°C and 1 atm, then:
   - Volume occupied by the oxygen:

$$V_{O_2} = \frac{m_{O_2} R_{O_2} T}{p} = \frac{8.29 \times 10^{-6} kg \times 0.2598 \, kJ \, / \, kg \cdot K \times 298.15 K}{3.17 \, kPa}$$

$$= 2.03 \times 10^{-4} m^3 = 0.203 \, L$$

   - Volume occupied by nitrogen:

$$V_{N_2} = \frac{m_{N_2} R_{N_2} T}{p} = \frac{13.9 \times 10^{-6} kg \times 0.2968 \, kJ \, / \, kg \cdot K \times 298.15 K}{3.17 \, kPa}$$

$$= 3.88 \times 10^{-4} m^3 = 0.388 \, L$$

- Vapor volume:

$$V_{vap} = \frac{m_v R_v T}{p} = \frac{0.997\,kg \times 0.4615kJ\,/\,kg \cdot K \times 298.15K}{3.17kPa}$$

$$= 43.276m^3 = 43,276L$$

- Oxygen concentration of vapor mixture:

$$C = \frac{V_{O_2}}{V_{mix}} = \frac{V_{O_2}}{V_{O_2} + V_{N_2} + V_{vap}} = \frac{0.203}{0.203 + 0.388 + 43276} = 4.69 \times 10^{-4}\%$$

It is to be noted that even though we did not consider any degassing intentionally, the resulting levels of oxygen concentration of heated steam and room-temperature vaporization are virtually the same and very low, of the order of 10^{-6} in terms of volume fraction. Once we consider the deaeration to be involved ahead of evaporation in both cases, the expected oxygen concentration of vapor mixture may reduce further by two to three orders of magnitude, considering the achievable deaeration is of the order of ppb. This is the potential both vapor bubbles and steam in deaeration process have in common.

So far, we have briefly introduced the generation conditions and composition information of vapor bubbles in relation to vapor bubble generation, which is the second step of vacuum bubbling. Vapor bubbles absorb dissolved oxygen in the solution through the mass diffusion process driven by the virtually zero oxygen concentration inside, realizing a very low dissolved oxygen concentration level. To prove this, a degassing experiment was conducted to check the performance of vacuum bubbling using a separate experimental device (see Sec. 2.1.3), and some of the results, such as the change in dissolved oxygen concentration during degassing, are presented in Figure 1.12. This is a part of the results of degassing performance tests under various conditions using the vacuum bubbling experimental device introduced in Sec. 2.3.2.4 as a result of the experiments conducted in the bubble laboratory of Kongju National University. The degassing experiment was performed for 400 L of fresh tap water at room temperature, and the dissolved oxygen concentration (mg/L) was converted from the measured dissolved oxygen concentration (% O_2 atm) using the saturated water vapor pressure at the corresponding temperature and Henry's constant. For reference, in this experiment, the internal pressure, p_1, of the tank is 1 kPa, and the depth of the bubbler is installed 0.3 m below the water surface.

According to the experimental results, the concentration of dissolved oxygen decreased very quickly at the beginning of the degassing process, and after this, the rate of decrease significantly slowed down. This can be seen as a result that proves

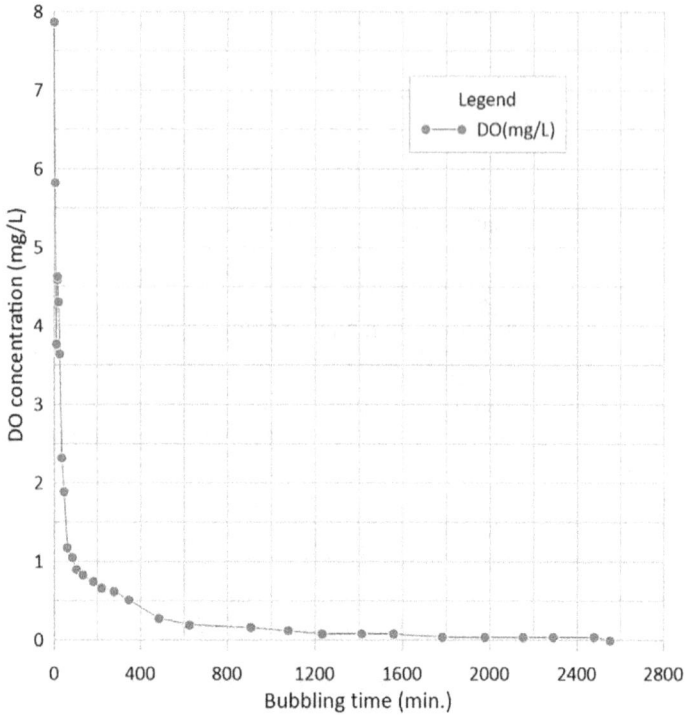

FIGURE 1.12 An example of measured DO concentration with vacuum bubbling deaeration.

the previous hypothesis that while the bubbles at the beginning of bubbling are supersaturated dissolved gases, once the supersaturated state is resolved to a certain extent, degassing occurs through mass diffusion by vapor bubbles. Specifically, the concentration of dissolved oxygen decreased very quickly from the initial 7.9 mg/L, reaching 1 mg/L after 88 min, 0.5 mg/L after 350 min, and 0 mg on the instrument after 2,551 min. Although more precise measurements at the ppb level were not made due to limitations in the precision of the measuring instrument, one of the most important results shown in this experiment is that deaeration of dissolved gases close to zero is possible through vacuum bubbling. Another important issue is energy consumption. Although the experimental condition applied in this test was not optimized at all in terms of efficiency or economics from the beginning, the amount of energy used during the degassing process was monitored to discuss energy consumption, which is an important issue in the degassing process. The electrical energy consumed for vacuum bubbling is the sum of the electricity consumption of the water pump for operating the bubbler and the vacuum pump used to depressurize the container. The input capacity of the water pump was fixed at a nominal 20 W and was calculated by multiplying the bubbling time, and the electricity consumption of the vacuum pump was measured using an integrated power meter. As a result of this test, the total power consumption for 400 L of water from the initial dissolved oxygen concentration of

7.9 mg/L to 0 mg/L was found to be 1.67 kWh [16]. The results of this experiment are considered meaningful in that they show that a high level of degassing is possible through vacuum bubbling and that quantitative energy consumption can be presented. Efforts to realize optimum operating conditions are still ongoing as the manuscript is being written, and we hope that more researchers will contribute to the development of this field with better results in the future.

So far, I have introduced deaeration as an application example of two-stage vacuum bubbling. However, in order to actually realize a low-energy, high-performance degassing device, research on not only the generation of vapor bubbles but also many design variables appears to be necessary. This may include the amount of bubbles generated, the size of the bubbles and their distribution, the location of the bubble generation (depth in the water), the temperature of the water, the shape of the bubbler nozzle, and the capacity of the pump. Chapter 2 covers physical phenomena related to vacuum bubbling and more technical details for system configuration and optimization.

The final topic deals with an introduction to desalination, which is considered a more aggressive application of vacuum bubbling. According to thermodynamic theory, the creation of vapor bubbles is ultimately a product of the phase change of the material, and the energy required for the phase change of the material, especially when a liquid vaporizes, has already been documented. (See Figure 1.13.)

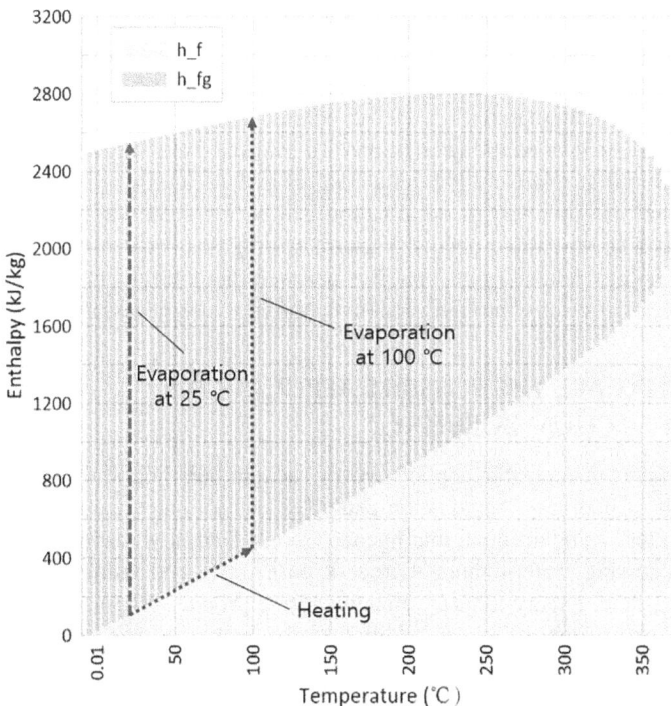

FIGURE 1.13 Enthalpy of saturated water [19].

Evaporative desalination is basically a process of heating water (seawater) to evapo-rate it and then condensing the evaporated vapor through a cooling device to obtain fresh water. Although various desalination technologies have been proposed and are currently in use so far, we would like to talk about the advantages that can be gained by realizing a large amount of vapor bubbles or evaporation under low vacuum, which is the third stage of vacuum bubbling.

Figure 1.13 is the data showing saturated water and vaporization enthalpy (h_{fg}) from the water temperature of 0°C to the critical temperature of 373.95°C. This data shows that the energy required to vaporize water is greater at lower temperatures. In particular, if you consider the temperature range of interest from 0°C to 100°C, you can see that the proportion of vaporization energy is much larger than the energy required to increase temperature. Now, let us compare the heat energy required for complete evaporation when water is heated from an initial temperature of 25°C to 100°C under atmospheric pressure and then evaporated (case 1) and when water is evaporated at room temperature 25°C (case 2). Of course, the pressure in the two cases should be different: atmospheric pressure and vacuum. At an initial tempera-ture of 25°C, the enthalpy of evaporation is 2,546.5 kJ/kg, and the enthalpy change when evaporating after heating to 100°C is 2,675.6 kJ/kg. The difference is about 5%, so the amount of energy required for evaporation does not vary significantly depending on temperature. The big difference comes not from the amount of vapor-ization energy but from the energy source. In order to create and maintain high-tem-perature conditions of 100°C or higher, most rely on fossil fuels. Since the thermal efficiency of conventional boilers is less than 40% [25], this means that we must con-sume more than twice as much fossil fuel as the heat energy we will actually use. On the other hand, when evaporating at a low temperature using a vacuum, renewable energy sources rather than fossil fuels can be used when slight heating or pumping is required to create vaporization conditions. This means that if large-scale evaporation can be achieved at relatively low temperatures in a vacuum, the enormous amount of fossil fuel use involved in desalination can be replaced with low-level energy sources, such as renewable energy. Vacuum heating has been introduced as a pub-lished technology related to this, but it is believed that a technical review of a large vapor generation system through vacuum bubbling is also necessary.

1.13 WHAT ARE THE POSSIBLE APPLICATIONS OF VACUUM BUBBLING?

There are many conceivable applications of vacuum bubbling at room temperature. As a result of querying "application examples of room-temperature vacuum bub-bling of water" using artificial intelligence Google Bad, it was found that it is used in food processing, water treatment process, purification and disinfection of water in the medical field, energy industry, manufacturing product processing, desalination, degassing, etc. Among these, in addition to the first-stage vacuum bubbling field that utilizes the pressure dependence of solubility, there are also cases using vapor bubble generation (second stage) and massive evaporation (third stage). Although it is not possible to present all the applications of vacuum bubbling, I personally would like to present four challenging application areas in this book. They are (1) process degassing; (2) underwater breathing, or so-called "artificial gills"—extracting high-oxygen air from water for respiration; (3) degassing of jet fuel; and (4) desalination. In the case of (2) and (3), it can be seen as the application fields of the first phase of vacuum bubbling, that is, using the pressure dependence of solubility, and (1) uses both the first and the second phases of vacuum bubbling using mass diffusion by vapor bubbles, and (4) can be seen as an application of the third phase of vacuum bubbling in which vacuum bubbling is realized in large quantities. Now, more specific physical phenomena will be discussed in Chapter 2, and application concepts for individual applications will be introduced in Chapter 3.

REFERENCES

[1] Y. D. Jun, "Energy field challenges to respond to climate change – focusing on IRENA's World Energy Transitions Outlook 2022," Newsletter of the Korean Society for New and Renewable Energy, May 2023.

[2] Ministry of Foreign Affairs of the Republic of Korea, (Joint Press Release) "Intergovernmental panel on climate change, approval of the 6th assessment report," March 20, 2023. www.mofa.go.kr/www/brd/m_4080/view.do?seq=373483

[3] KBS, [Issue at a glance] " 'Decision within 10 years of earth's existence' UN's dire warning . . . our response?," March 22, 2023. www.youtube.com/watch?v=aGg-Z1X9v9g

[4] IRENA, World Energy Transitions Outlook 2022: 1.5°C Pathway, Executive Summary of IRENA. International Renewable Energy Agency, Abu Dhabi, 2022. www.irena.org/publications

[5] Deaeration Machine Market: Global Industry Analysis and Forecast (2021–2027), by Type, Function, Industry, and Region. Maximize Market Research Pvt. Ltd. www.maximizemarketresearch.com/market-report/global-deaeration-machine-market/88196/

[6] The Engineering ToolBox, "Solubility of air in water," 2004. www.engineeringtoolbox.com/air-solubility-water-d_639.html. Retrieved August 25, 2023.

[7] J. W. Lee, P. W. Heo, T. S. Kim, "Theoretical model and experimental validation for underwater oxygen extraction for realizing artificial gills," Sensors and Actuators A: Physical, 2018, 284, pp. 103–111.

[8] I. Ieropoulos, C. Melhuish, J. Greenman, "Artificial gills for robots: MFC behavior in water," Bioinspiration & Biomimetics, 2007, 2, pp. S83–S93.

[9] A. Bodner. www.likeafish.biz. Retrieved November 3, 2023.

[10] G. B. Kim, S. J. Kim, M. H. Kim, C. U. Hong, H. S. Kang, "Development of a hollow fiber membrane module for using implantable artificial lung," Journal of Membrane Science, 2009, 326, pp. 130–136.

[11] G. R. Greenbank, P. A. Wright, "The deaeration of raw whole milk before heat treatment as a factor in retarding the development of the tallowy flavor in its dried product," Journal of Dairy Science, 1951, 34(8), pp. 815–818.

[12] H. Carlsson, C. Jonsson, "Separation of air bubbles from milk in a deaeration process," May 2012. www.researchgate.net/publication/267687974

[13] "Water treatment: Aeration and gas stripping—TU Delft OpenCourseWare." https://ocw.tudelft.nl/제-content/uploads/Aeration0and-gas-stripping-1.pdf. Retrieved December 24, 2021.

[14] D. Arthur, "How does deaerator work?," September 22, 2019. http://watertreatmentbasics.com/how-does-deaerator-work/. Retrieved December 24, 2021.

[15] Y.-D. Jun, "Degassing dissolved oxygen through bubbling: The contribution and control of vapor bubbles," Processes, 2023, 11(11), p. 3158. https://doi.org/10.3390/pr11113158.

[16] S. H. Yoo (ed.), Encyclopedia of Soil Science. Seoul National University Press, Seoul, 2000, pp. 443–444. ISBN 89-521-0204-5. (In Korean)

[17] J. A. Hong, J. S. Lee, Y. D. Jun, "Degassing dissolved oxygen through bubbles under a vacuum condition," Proceedings of 7th Thermal and Fluids Engineering Conference (TFEC) Held at UNLV, Las Vegas, NV, April 2022. Hosted by American Society of Thermal and Fluids Engineers, pp. 1021–1033, TFEC-2022-41702.

[18] J. Saleh (ed.), Fluid Flow Handbook. McGraw-Hill, New York, NY, 2002.

[19] Y. A. Cengel, M. A. Boles, Thermodynamics—An Engineering Approach, 5th ed. in SI Units. McGraw-Hill, New York, NY, 2006, p. 890.

[20] U.S. Department of Energy, "Energy Efficiency and Renewable Energy, Steam System Modeler Tool (SSMT)." https://www4.eere.energy.gov/manufacturing/tech_deployment/amo_steam_tool/overview. Retrieved November 9, 2023.

[21] H. Hu, C. Xu, Y. Zhao, K. J. Ziegler, J. N. Chung, "Boiling and quenching heat transfer advancement by nanoscale surface modification," Scientific Reports, 2017, 7, p. 6117. https://doi.org/10.1038/s41598-017-06050-0; www.nature.com/scientificrepoerts.

[22] A. Prosperetti, "Vapor bubbles," Annual Review of Fluid Mechanics, 2017, 49, pp. 221–248. http://doi.org/10.1146/annurev-fluid-010816-060221

[23] C. E. Brennen, Cavitation and Bubble Dynamics. Oxford University Press, New York, 1995.

[24] P. Chiggiato, "Vacuum technology for superconducting devices," Proceedings of the CAS-CERN Accelerator School: Superconductivity for Accelerators, Erice, April 24–May 4, 2013, edited by R. Bailey, CERN-2014-005. CERN, Geneva, 2014.

[25] P. Basu, C. Kefa, L. Jestin, Boilers & Burners: Design & Theory (Korean Version). KEIC, Seoul, Korea, 2000, p. 26.

2 Solubility, Diffusion, and Evaporation

The bubble generation mechanism in vacuum is thought to be contributed by the solubility limit set by the thermodynamic state and the phase change of pure substances such as water. Once the solubility limit is lowered by having a lowered pressure, some of the solute already present becomes supersaturated and is placed under conditions that make it difficult to stay in the dissolved form in water system any longer. When this situation occurs inside a liquid, the supersaturated solute gas separates from the solution in the form of bubbles. The dependence of solubility on temperature and pressure is well established, and exploiting this dependence of solubility on temperature was a natural choice for the power industry, where heating water and handling high-pressure steam are part of the process. If we look at other options than heating, we can think of depressurization as a way to reduce solubility without resorting to heating. One of the most immediate driving forces for obtaining a bubble in this case arises from the dependence of solubility on pressure according to Henry's law. As the pressure inside the container with liquid decreases, more and more of the dissolved gases originally dissolved in the liquid solvent are situated in a supersaturated condition and ready to escape from the liquid to reach equilibrium. In this case, the interior of the bubble consists of supersaturated solute gas components and saturated water vapor. The barriers that limit process transition from supersaturated to saturated equilibrium condition seem to be minimized by resorting to an arbitrary way of mixing. Once this process initiates, the resulting phenomena becomes any kind of bubble generation. However, since solubility is typically defined for solvents in a liquid state, once the pressure reaches down its saturated vapor pressure, where phase change from liquid to vapor may occur, then Henry's law becomes no more valid. This defines the application limit of solubility-driven deaeration. For example, water at 25°C may no longer exist as liquid water when the pressure reaches its saturated vapor pressure of 3.17 kPa and can change into gaseous vapor, so the limit value of solubility is 0.27 mg/L ($G_{O_2} = K_{O_2} p_{O_2} = 1.28 \times 10^{-3} mole / \left(atm \cdot L \right) \times \dfrac{3.17}{101.3} \times 0.209$ $atm \times 32g / mole \times 1000mg / g = 0.27mg / L$). The next step in bubbling is vaporization and the end result is vapor bubbling. Vapor bubble generation has been considered to be a challenge in that the pressure must be maintained at least equal to or below the saturated vapor pressure; however, given the current advances in vacuum technology, this is not that difficult. The composition of the vapor bubble is essentially water vapor, and the dissolved gas concentration is very low (near zero) due to the volume expansion during the phase-change process. Due to the generation of vapor bubbles, deaeration can be processed further below the solubility limit set by the phase change of solution. Degassing through vapor bubbles is achieved through the mass diffusion process by the concentration

DOI: 10.1201/9781003374626-2

difference between the dissolved gas in the water and the inside of the vapor bubbles, and Fick's diffusion law or its modified form can be applied here. In this process, the size and number of bubbles and the length of stay in liquid according to the depth of bubble formation will have a direct effect on mass diffusion. The size of the bubble in the boiling process is known to be a function of the degree of superheating [1], but not clearly understood in the case of depressurization (tension) process. The generation of bubbles by these two mechanisms is, however, difficult to distinguish in practice, because we only observe bubbles that form when the pressure is kept low. According to a series of experiments to clarify this point, the composition of bubbles generated under conditions where the level of vacuum is higher than the saturated vapor pressure is similar to the composition of dissolved gas, and the oxygen concentration is as high as 32% [2]. It was confirmed through experiments that the oxygen concentration of the captured gas decreases when the vacuum level is lowered below the vapor pressure. If the vapor bubble generation mechanism is clear, it will be possible to present another version of the currently widely used evaporative desalination process by generating and condensing steam bubbles in large quantities at room temperature.

2.1 SOLUBILITY

2.1.1 GENERAL BEHAVIOR OF SOLUBILITY: PRESSURE AND TEMPERATURE DEPENDENCY

According to the Wikipedia Dictionary on *solubility* [3], the *solubility* in chemistry is defined as the ability of a substance, a solute, to form a solution with another substance, a solvent. It is measured as the concentration of solute in a saturated solution at which no more solute can be dissolved [4]. At this point, the two substances are said to be in solubility equilibrium. Solubility depends primarily on the composition of the solute and solvent (including their pH and the presence of other dissolved substances), as well as on temperature and pressure. Under certain conditions, the concentration of a solute can exceed its usual solubility limit, resulting in a supersaturated solution, which is known to be metastable and rapidly reject excess solute when suitable nucleation sites appear [5].

Solubility decreases with temperature, and the case of water solubility at 1 atm is illustrated in Figure 2.1, while it is known that the pressure-dependent behavior of solubility is governed by Henry's law. Henry's law can be expressed in various forms depending on the application field (see sidebar #1), and this book introduces it as follows:

The concentration (mol/L) of a dissolved gas that can be dissolved in a solution in equilibrium is proportional to the partial pressure of the gas, and the proportionality constant at this time is called Henry's solubility law constant, or simply Henry's constant [7]. One form of Henry's law states that:

$$C_s = Hp_i \tag{1}$$

where C_s is the solubility or equilibrium concentration (mol/L), H is Henry's solubility law constant (mol/L·atm), and p_i is the partial pressure of a specific gas (atm). Figure 2.2 illustrates the solubility behavior of oxygen and nitrogen in water at 25°C according to Henry's law, with the constant values shown in Table 2.1.

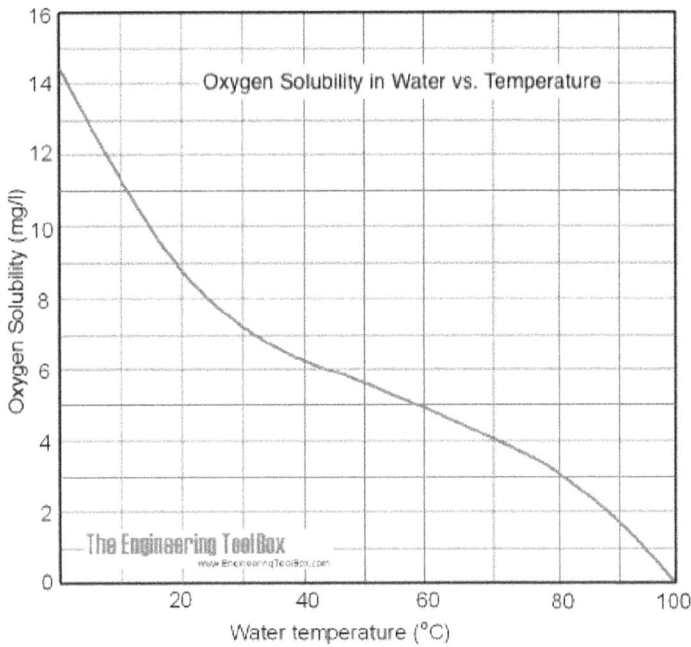

FIGURE 2.1   Oxygen solubility in fresh water [6].

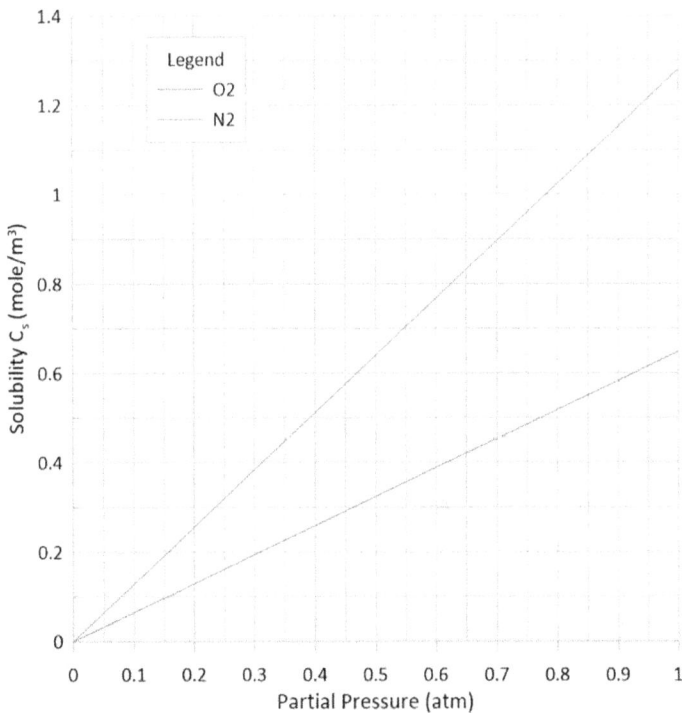

FIGURE 2.2   Solubility of oxygen and nitrogen in water at different pressure at 25°C.

TABLE 2.1
Henry's Solubility Law Constants for Selected Gases in Water at 25°C

Henry's Constant	Unit	Value
H_{O_2}	mol/(L·atm)	1.28×10^{-3}
H_{N_2}		6.48×10^{-4}
H_{CO_2}		3.38×10^{-2}
H_{H_2}		7.90×10^{-4}
H_{CH_4}		1.34×10^{-3}
H_{NO}		2.0×10^{-3}

Source: [7].

Since the solubility of gases in the atmosphere also involves the partial pressure of water vapor itself, it is necessary to correct the partial pressure according to the temperature. For example, when calculating the solubility of saturated oxygen in water under an atmospheric pressure of 1 atm, the partial pressure of water vapor at 25°C is 0.0313 atm, and oxygen occupies 20.95% of the assumed dry air, so the partial pressure of oxygen (p_{O_2}) would be (1.00 − 0.0313) atm × 0.2095 = 0.2029 atm. Therefore, according to Henry's law, the concentration of dissolved oxygen is calculated as follows [7]:

$$C_{s,O_2} = H_{O_2} p_{O_2} = 1.28 \times 10^{-3} mole \cdot L^{-1} \cdot atm^{-1} \times 0.2029 \, atm$$
$$= 2.60 \times 10^{-4} mole \cdot L^{-1} = 2.60 \times 10^{-4} \times 32 \times 10^{3} mg \cdot L^{-1} = 8.31 mg \cdot L^{-1}$$

Noting that the Earth's atmosphere is mainly composed of nitrogen, oxygen, and other gases, as shown in Table 2.2, it can be assumed that the air is composed of 79.1% nitrogen and 20.9% oxygen for the convenience of calculation.

TABLE 2.2
Constituents of Atmosphere

Constituents	Fraction (%)
Nitrogen (N_2)	78.084
Oxygen (O_2)	20.947
Argon (Ar)	0.934
Carbon dioxide(CO_2)	0.035

Source: Quoted from [8].

If the pressure is lowered in an idealized atmosphere but remains above the saturated vapor pressure, the bubbles extracted will have different composition from the standard air. For water initially under concentration equilibrium with an idealized atmosphere at 25°C and 1 atm, the dissolved oxygen and nitrogen concentrations can be obtained from the following equations:

$$C_{s,O_2} = H_{O_2} P_{O_2} \left(\frac{mol}{L} \right) = H_{O_2} P_m \left(1 - C_{vap} \right) C_{O_2} M_{O_2} \times 1000 \left(\frac{mg}{L} \right) \tag{2}$$

$$C_{s,N_2} = H_{N_2} P_{N_2} \left(\frac{mol}{L} \right) = H_{N_2} P_m \left(1 - C_{vap} \right) C_{N_2} M_{N_2} \times 1000 \left(\frac{mg}{L} \right) \tag{3}$$

where P_m is the pressure of mixture air, C_{vap} the volume fraction of vapor in a saturated condition with $C_{O_2} = 0.209$, $C_{N_2} = 0.791$ and $M_{O_2} = 32g/mol, M_{N_2} = 28g/mol$, respectively.

Let us consider a case where the pressure around water, which was initially in equilibrium with atmospheric pressure of 25°C, is lowered to 0.1 atm and reaches equilibrium (Figure 2.3). Remember that the saturated vapor pressure of water at 25°C is 3.17 kPa, so 0.1 atm (10.13 kPa) is still higher than the saturated vapor pressure. The amount of gas that can be extracted through bubbles per liter of water

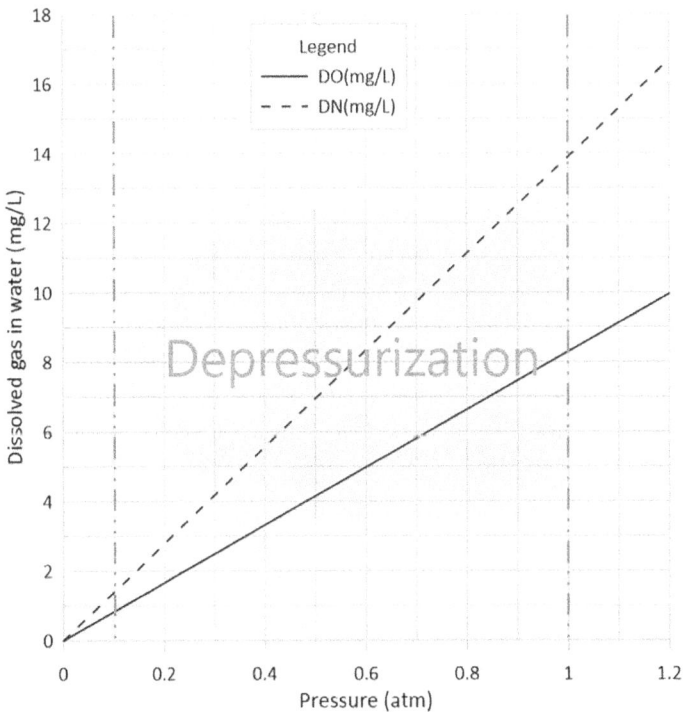

FIGURE 2.3 Expected dissolved gas contents (mg/L) of oxygen and nitrogen in water at 25°C at lowered pressure under an equilibrium state.

due to the solubility limit is 2.33×10^{-4} moles for oxygen and 4.47×10^{-4} moles for nitrogen, according to equations (2) and (3). Considering that the volume occupied by 1 mole of gas at 25°C and 1 atm is $22.4L \times \dfrac{298.15}{273.15} = 24.45L$, it will be possible to capture $(2.33 + 4.47) \times 10^{-4} \, mol \times 24.45 \, L \, / \, mole = 0.01663L$ of extract gas (0.0057 L of oxygen, 0.01093 L of nitrogen) out of 1 L of water. In this case, the oxygen volume fraction of non-condensable gases (oxygen and nitrogen), excluding the contribution of water vapor, is expected to be $\dfrac{0.0057}{0.0057 + 0.1093} = 0.343$, that is, 34.3%.

Here, to minimize complexity, we have talked about water in equilibrium with idealized air (composed of 20.9% oxygen and 79.1% nitrogen) at 25°C, but in reality, the temperature of water and the pressure conditions of the atmosphere can vary, and due considerations should be paid for each condition. Of particular interest is that the saturated vapor pressure of water is a function of temperature only and exhibits a constant partial pressure regardless of depressurization, which means that the relative proportion of vapor pressure increases as the pressure decreases. This discussion continues in Sec. 2.2.

Another issue related to solubility is the limit of solubility. Previously, we looked at an example where the ambient pressure of water initially in equilibrium under atmospheric pressure decreased to 0.1 atm. In this case, under the assumption that the initial composition of the mixed air remains the same, it can be said that the solubility of dissolved gases decreased by 90%. That portion of the supersaturated solute is unstable and can be removed by mixing. However, Henry's solubility law is no longer applicable when the solvent undergoes a phase change from liquid to vapor at low pressures near the saturated vapor pressure. When the pressure of water falls below its saturated vapor pressure, water may no longer exist in a stable liquid state.

SIDEBAR 1: QUANTIFICATION OF SOLUBILITY (QUOTED FROM SOLUBILITY, WIKIPEDIA) [9]

There are different ways to express solubility, the concentration of a solution, such as the mass, volume, or moles of solute relative to a specific mass, volume, or molar amount of solvent or solution. For solutions of liquids and gases in liquids, the quantities of both substances may be given volume rather than mass or mole amount, such as liter of solute per liter of solvent, or liter of solute per liter of solution. The value may be given as a percentage, and the abbreviation "v/v" for "volume per volume" may be used to indicate this choice. According to [10], the solubility of oxygen in water is about 0.0013 M (mole/L) at 1 bar, and the vapor pressure of a gas in the equilibrium vapor phase in the low-mole-fraction region is proportional to the mole fraction of this gas. In other words:

$$p_i = K_i x_i \tag{a}$$

where the value of the proportionality constant, K_i, depends on the component and temperature. This equation is called Henry's law, named after British chemist William Henry, and K_i is called Henry's law constant. For example, suppose the Henry's law constant for oxygen in water at 25°C is $4.34 \times 10^9 \, Pa$, and the equilibrium partial pressure of oxygen at the same temperature is $p_{O_2} = 1 bar = 10^5 \, Pa$, then what is the mole fraction of oxygen in this aqueous solution? From the preceding relation:

$$10^5 \, Pa = 4.34 \times 10^9 \, Pa \times x_{O_2}$$

$$x_{O_2} = \frac{10^5 \, Pa}{4.34 \times 10^9 \, Pa} = 2.304 \times 10^{-5} \, \frac{mol - O_2}{mol - water}$$

The mass of 1 mole of water can be seen as 18 g (molecular weight of 18), and the density of water at 25°C is 997 kg/m³, so 1 L of water is equivalent to 997/18 = 55.39 mole, or 0.018 L/mole. Therefore, the molarity (mole/L) becomes:

$$2.304 \times 10^{-5} \, \frac{mol - O_2}{mol - water} \times \frac{1 mol - water}{0.018 L - water}$$

$$= 0.00128 \, \frac{mol - O_2}{L - water} = 0.00128 \, \frac{mol}{L}$$

One can use this kind of relationship to convert one form of solubility to another in the literature.

TABLE 2.A

Henry's Law Constants for Several Aqueous Solutions at 25°C

Compound	$K_i \, (Pa)$
Argon (Ar)	4.03×10^9
Carbon Dioxide (CO_2)	1.67×10^8
Nitrogen (N_2)	8.57×10^9
Oxygen (O_2)	4.34×10^9

Source: Partial excerpt from Table 7.2 from [10].

SIDEBAR 2: HENRY'S LAW

Henry's law is a gas law that states that the amount of gas dissolved in a liquid is proportional to the partial pressure above the liquid [11]. The constant of proportionality in the law is called Henry's law constants, and there are several ways to define them which can be subdivided into two basic types. One possibility is to put the aqueous phase into the numerator and the gaseous phase into the denominator ("aq/gas"). This results in the Henry's law solubility constant H_S. Its value increases with increased solubility. Alternatively, numerator and denominator can be switched ("gas/aq"), which results in the Henry's law volatility constant H_V. The value of H_V decreases with increased solubility. International Union of Pure and Applied Chemistry (IUPAC) describes several variants of both fundamental types [12], which result from the multiplicity of quantities that can be chosen to describe the composition of two phases. Typical choices for the aqueous phase are molar concentration (c_a), molality (b), and molar mixing ratio (x). For the gas phase, molar concentration, c_g, and partial pressure, p, are often used. To specify the exact variant of the Henry's law constant, two superscripts are used. They refer to the numerator and the denominator of the definition. For example, H_S^{cp} refers to the Henry solubility, defined as c / p. Henry's law solubility constant, which is mainly used in this book, is defined on the basis of solubility as follows:

$$H_S^{cp} = \frac{c_a}{p}$$ (b)

where c_a is the concentration of a species in the aqueous phase, and p is the partial pressure of that species in the gas phase under equilibrium conditions. The SI unit for H_S^{cp} is mol/(m³·Pa); however, often, the unit M/atm is used, since c_a is usually expressed in M (1 M = 1 mol/dm³) and p in atm (1 atm = 101325 Pa). In addition, Henry's solubility (H_S^{cp}), defined through the aqueous-phase mixing ratio, and Henry's solubility (H_S^{bp}), defined through the molality, are also used, and these values are mutually convertible through a conversion process. Figure 2.A shows some selected gas data (an image) among the Henry's constant data announced by Sander (2015). [13]

Gas	$H_V^{pc} = \dfrac{p}{c_{aq}}$ $\left(\dfrac{L \cdot atm}{mol}\right)$	$H_S^{cp} = \dfrac{c_{aq}}{p}$ $\left(\dfrac{mol}{L \cdot atm}\right)$	$H_V^{px} = \dfrac{p}{x}$ (atm)	$H_S^{cc} = \dfrac{c_{aq}}{c_{gas}}$ (dimensionless)
O_2	770	1.3×10^{-3}	4.3×10^4	3.2×10^{-2}
H_2	1300	7.8×10^{-4}	7.1×10^4	1.9×10^{-2}
CO_2	29	3.4×10^{-2}	1.6×10^3	8.3×10^{-1}
N_2	1600	6.1×10^{-4}	9.1×10^4	1.5×10^{-2}
He	2700	3.7×10^{-4}	1.5×10^5	9.1×10^{-3}
Ne	2200	4.5×10^{-4}	1.2×10^5	1.1×10^{-2}
Ar	710	1.4×10^{-3}	4.0×10^4	3.4×10^{-2}
CO	1100	9.5×10^{-4}	5.8×10^4	2.3×10^{-2}

FIGURE 2.A   Henry's law constants (gases in water at 298.15K) (image) [13].

SIDEBAR 3: SOLUBILITY OF OXYGEN IN WATER WITH RESPECT TO TEMPERATURE

The solubility of oxygen in water as a function of temperature is of great interest, and numerous researchers have reported related results. The contents introduced here are selected from literature in various study fields. The latest engineering simulation tools should have these data converted into a database and prepared in a usable form.

Sander (2015) [13] introduced the behavior of Henry's law constants according to the van't Hoff equation, which explains the temperature dependence of equilibrium constants. In other words, when the temperature of the system changes, the Henry constant behaves as follows:

$$\frac{dlnH}{d\left(\frac{1}{T}\right)} = \frac{-\Delta_{sol}H}{R} \tag{c}$$

where Δ_{sol} is the enthalpy of dissolution. When the equation is integrated, referring to the reference value of H^0 at the reference temperature $T^0 = 298.15K$, it yields [13]:

$$H(T) = H^0 exp\left[\frac{-\Delta_{sol}H}{R}\left(\frac{1}{T} - \frac{1}{T^0}\right)\right] \tag{d}$$

As the temperature is raised, gases usually become less soluble in water. He also reported a compilation of Henry's law constants for water as solvent, in which the data for oxygen in terms of H^{cp} at $T^\ominus = 298.15K$ and $\frac{dlnH^{cp}}{d\left(\frac{1}{T}\right)}$ are reported, as quoted in Figure 2.B.

Fernández-Prini et al. [14] reported correlations for the Henry's constant k_H for the solutes in H_2O and D_2O which commonly might be encountered in geochemistry or the power industry. Interestingly, their collected data for oxygen covered a wide range of temperature from 274.15 K to 616.52 K. Here, Henry's constants were fitted to the equation proposed by Harvey (1996) [15]:

$$ln\left(k_H / p_1^*\right) = \frac{A}{T_R} + \frac{B\tau^{0.355}}{T_R} + C\left(T_R\right)^{-0.41} exp\tau, \tag{e}$$

where $\tau = (1-T_R)$, $T_R = T/T_{cl}$, T_{cl} is the critical temperature of the solvent as accepted by IAPWS [16] (647.096 K for H_2O), and p_1^* is the vapor pressure of the solvent at the temperature of interest. Those parameter values for

Table 6: Henry's law constants for water as solvent

Substance Formula (Other name(s)) [CAS registry number]	H^{cp} (at $T^{\ominus}$) $\left[\dfrac{mol}{m^3\,Pa}\right]$	$\dfrac{d\ln H^{cp}}{d(1/T)}$ [K]	Reference	Type	Note
Inorganic species					
Oxygen (O)					
oxygen	1.2×10^{-5}	1700	Warneck and Williams (2012)	L	
O_2	1.3×10^{-5}	1500	Sander et al. (2011)	L	
[7782-44-7]	1.3×10^{-5}	1500	Sander et al. (2006)	L	
	1.3×10^{-5}	1400	Fernández-Prini et al. (2003)	L	1
	1.3×10^{-5}	1500	Battino et al. (1983)	L	
	1.3×10^{-5}	1500	Wilhelm et al. (1977)	L	
	1.3×10^{-5}	1400	Rettich et al. (1981)	M	
	1.3×10^{-5}	1400	Benson et al. (1979)	M	
	1.2×10^{-5}	1800	Carpenter (1966)	M	
	1.3×10^{-5}	1200	Winkler (1891b)	M	2
	1.3×10^{-5}	1500	Battino (1981)	X	3, 4
	1.3×10^{-5}	1500	Battino (1981)	X	5
	1.2×10^{-5}	1700	Dean (1992)	?	6
	1.3×10^{-5}		Seinfeld (1986)	?	7

FIGURE 2.B   Collected tabulated data of H^{cp} at $T^{\ominus} = 298.15K$ and $\dfrac{d\ln H^{cp}}{d\left(\dfrac{1}{T}\right)}$ (partly quoted from [13]).

(Calculated from data in Table 9.1 and Henry's law)

Temperature (°C)	Oxygen solubility (kg m^{-3})
0	1.48×10^{-2}
10	1.15×10^{-2}
15	1.04×10^{-2}
20	9.45×10^{-3}
25	8.69×10^{-3}
26	8.55×10^{-3}
27	8.42×10^{-3}
28	8.29×10^{-3}
29	8.17×10^{-3}
30	8.05×10^{-3}
35	7.52×10^{-3}
40	7.07×10^{-3}

FIGURE 2.C   Solubility of oxygen in water under 1 atm air pressure (data image) [17].

oxygen are: A = -9.44833, B = 4.43822, C = 11.42005, $T_{min}(K) = 274.15$, and $T_{max}(K) = 616.52$.

Doran [17] introduced the variation of oxygen solubility with temperature for water in the range 0 to 40°C (Figure 2.C) under 1 atm air pressure for applications in bioprocess engineering. Oxygen solubility falls with increasing

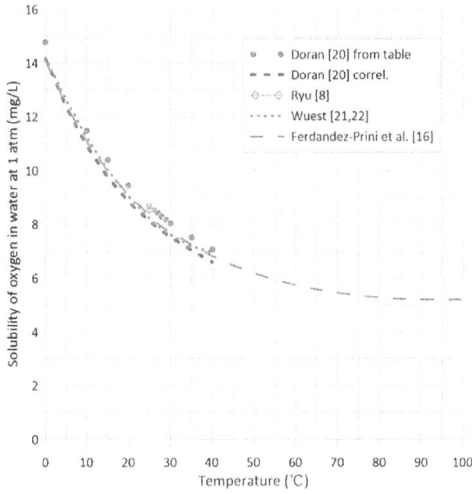

FIGURE 2.D   Compiled correlated data for oxygen solubility in water with temperature.

temperature, and the solubility of oxygen from air in pure water between 0°C and 36°C has been correlated using the following equation [17]:

$$C^* = 14.161 - 0.3943T + 0.007714T^2 - 0.0000646T^3 \tag{f}$$

where C^* is oxygen solubility in units of mg/L, and T is temperature in °C. Wuest et al. [18] and McGinnis et al. [19] presented the solubility constants of oxygen and nitrogen in water as a function of temperature in water treatment applications. Unfortunately, the applicable temperature range of this correlation was not specified.

$$H_O = 2.125 - 5.021 \times 10^{-2}T + 5.77 \times 10^{-4}T^2 \text{ (T in°C)} \tag{g}$$

$$H_N = 1.042 - 2.450 \times 10^{-2}T + 3.171 \times 10^{-4}T^2 \text{ (T in°C)} \tag{h}$$

Figure 2.D is obtained from data on the solubility of oxygen in water according to temperature changes collected from the referred literature. It can be seen that the solubility behavior of oxygen in water shows good agreement with the general behavior of decreasing with increasing temperature from 0°C to 40°C, the temperature range for which actual measurement data are available. Among them, Fernández-Prini et al. (2003) are notable in that they provide data for a temperature range from 1°C to 343°C, whereas in this figure, only up to 100°C is presented. According to this data, even when the temperature reaches 100°C, the solubility is about 5 mg/L, which is different from the results presented in other literature depending on temperature, so caution needs to be paid when using the data. The collected data suggests that the solubility of oxygen has a maximum value near 0°C and decreases to about

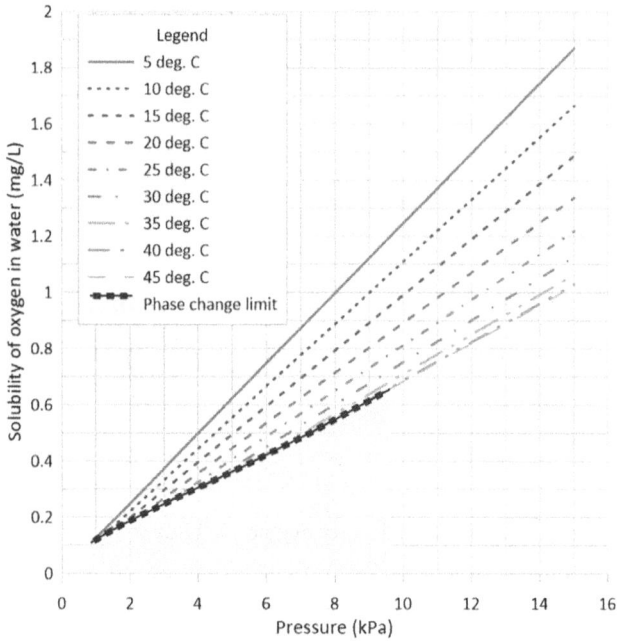

FIGURE 2.4   Solubility of oxygen in water under 1 atm with phase change limits.

half of the maximum value near 40°C, the highest temperature the solubility data is measured [17].

2.1.2   APPLICATION LIMIT OF SOLUBILITY

Another important issue related to solubility is its application limit. A material undergoes a phase change according to its thermodynamic state, also known in the form of phase change diagram. Henry's law deals with the saturated solubility of a solute in a liquid solvent, but if the pressure is so low that it reaches below the saturated vapor pressure of the liquid solvent, the phase of the solvent may change from liquid to vapor, in which case Henry's law on solubility is no longer applicable. Figure 2.4 is a solubility diagram considering the pressure limit at which the phase change of water can occur at various temperatures, that is, the saturation vapor pressure, as the lower limit for applying the solubility law. That is, the range of application of Henry's law is effective only within the range where the solution is under a pressure above the saturated vapor pressure or the solution exists in the liquid phase. Then, is it impossible to realize a lower solubility in water?

In the sidebar of his review paper (2017) [20], Prosperetti describes the role of dissolved permanent gas in the process of condensation of vapor bubbles generated by the heating method in contact with the surrounding subcooled liquid. In his article, the claims that vapor bubbles do not disappear or that vapor bubbles can be created without heating the liquid were dismissed as absurd claims. It seems that no researcher has officially refuted his declaration since then, but I am now claiming that room-temperature vapor bubbles can be created and maintained without heating. According to his argument, when the vapor bubbles created by heating come into contact with the surrounding subcooled liquid, the vapor condenses and the remaining permanent gas forms bubbles, which are sometimes misunderstood as vapor bubbles. However, in order to create vapor bubbles without heating the liquid, it is possible to generate vapor bubbles by placing the liquid locally at a sufficiently low pressure condition below the saturation vapor pressure, and by continuously providing conditions in which the generated vapor bubbles can be maintained, in which vapor bubbles can remain or grow. This is my argument, and it is also the core content of this book. Before discussing the second topic of vacuum bubbling, which is the generation and maintenance of vapor bubbles, it would be good to finish talking about other interesting topics related to solubility. So far, for the sake of a more general discussion, we have not specified the fluid of interest, but water is probably one of the most important fluids for vacuum bubbling applications. One of the things I am curious about is how much gas can be removed from water, what is the composition of the removed gas, and finally, in a fluid such as fuel, what is its solubility?

2.1.3   INTEREST IN COLLECTED GAS FROM DEGASSING

In most industrial applications, degassing of dissolved gases, especially dissolved oxygen in a liquid, is of major interest, but there are also areas of interest in the utilization of deaerated gases due to its interesting characteristics. A representative example is the use of degassed air from water, which has been entrenched in the so-called "artificial gills." The properties of the degassed air are, however, not only significant in applications such as "artificial gills." A more systematic approach is necessary for a better understanding of the outgassing phenomenon and for its engineering applications, especially in the design of deaeration systems. As discussed earlier, let us idealize the main components that make up the atmosphere to be composed of 79.1% nitrogen and 20.9% oxygen. Oxygen accounts for about 1/5 of the air, but its solubility is about twice that of nitrogen (Table 2.2). Therefore, based on this data, it is interesting to examine the composition of vapor and gas inside the bubble at 25°C and 1 atm. The partial pressures of saturated vapor, oxygen, and nitrogen in the atmosphere are, respectively:

$$\text{Water vapor}: \ p_{vap} = 3.17\,kPa \text{ or } 0.0313\,\text{atm}$$

$$\text{Oxygen}: \ p_{O_2} = 1\,\text{atm} \times (1 - 0.0313) \times 0.209 = 0.2025\,\text{atm}$$

$$\text{Nitrogen}: \ p_{N_2} = 1\,\text{atm} \times (1 - 0.0313) \times 0.791 = 0.7662\,\text{atm}$$

That is, noting that the volume fraction of oxygen in dry air is 20.9%, the volume fraction of oxygen of saturated air is 20.3%. And according to Henry's law, the concentrations of dissolved oxygen and dissolved nitrogen in 1 L of water are, respectively:

1. Oxygen

$$C_{s,O_2} = H_{O_2} p_{O_2} = 1.28 \times 10^{-3} \, mole \cdot L^{-1} \cdot atm^{-1} \times 0.2025 \, atm$$

$$= 2.59 \times 10^{-4} \, mole \cdot L^{-1} = 2.59 \times 10^{-4} \times 32 \times 10^3 \, mg \cdot L^{-1} = 8.29 \, mg \cdot L^{-1}$$

2. Nitrogen

$$C_{s,N_2} = H_{N_2} p_{N_2} = 6.48 \times 10^{-4} \, mole \cdot L^{-1} \cdot atm^{-1} \times 0.7662 \, atm$$

$$= 4.97 \times 10^{-4} \, mole \cdot L^{-1} = 4.97 \times 10^{-4} \times 28 \times 10^3 \, mg \cdot L^{-1} = 13.92 \, mg \cdot L^{-1}$$

And the volume occupied by this gas under an atmospheric pressure is:

1. Oxygen

$$V_{O_2} = \frac{m_{O_2} R_{O_2} T}{p} = \frac{8.29 \times 10^{-6} \, kg \times 0.2598 kJ \, / \, kg \cdot K \times 298.15K}{101.3 kPa}$$
$$= 6.34 \times 10^{-6} \, m^3 = 0.00634 L$$

2. Nitrogen

$$V_{N_2} = \frac{m_{N_2} R_{N_2} T}{p} = \frac{13.92 \times 10^{-6} \, kg \times 0.2968 kJ \, / \, kg \cdot K \times 298.15K}{101.3 kPa}$$
$$= 1.215 \times 10^{-5} \, m^3 = 0.0122 L$$

Therefore, the total amount of gas considering the contribution of water vapor becomes:

$$V = V_{O_2} + V_{N_2} + V_{vap} = 0.00634 + 0.0122 + V_{vap}.$$

Noting that in an ideal gas mixture the volume fractions equal to the mole fractions, we can write the following relation for vapor mole fraction as:

$$\frac{p_{vap}}{p} = \frac{p_{vap}}{p_{O_2} + p_{N_2} + p_{vap}} = \frac{V_{vap}}{V_{O_2} + V_{N_2} + V_{vap}},$$

And recalling that $\dfrac{p_{vap}}{p} = \dfrac{3.17}{101.3} = 0.0313$:

$$0.0313 = \frac{V_{vap}}{0.00634 + 0.0122 + V_{vap}}$$

From which $V_{vap} = 0.0006L$. In this case, the volume fraction or mole fraction of each component is $y_{O_2} = 0.331$, $y_{N_2} = 0.637$, and $y_{vap} = 0.031$. This approach, of course, can be extended for lower pressure values.

Through the preceding discussion, we can secure the basic data of the mass calculation in the bubble when the dissolved gas forms a bubble, and the volume fraction of oxygen under atmospheric pressure is about 33% (i.e., about 1/3 of the total volume), which is much higher than 20.9% in the standard atmosphere. In this way, by degassing and capturing the dissolved gas in the water, high-oxygen air having an oxygen concentration of about 1.5 times higher than that of the atmosphere can be obtained. However, it is worth noting that the amount of gas that can be extracted is limited to within 2% of the liquid amount under atmospheric pressure. This is believed to be a reference for researchers considering the direct application of artificial gills in estimating the amount of water that must be treated to obtain the amount of oxygen required for human respiration.

SIDEBAR 4: ON THE EQUILIBRIUM COMPOSITION INSIDE THE SOLUBILITY-DRIVEN BUBBLES

Assuming that the composition of dry air is 20.9% oxygen and 79.1% nitrogen, and that saturated water vapor exists, the composition of bubbles created under 1 atm was discussed in the text. Then, what will be the composition of the bubbles that are created when the surrounding pressure is lowered? Can the oxygen volume fraction of 33.1% always be maintained?

This question arose from the results obtained during experiments with actual pressure changes. This is because the oxygen concentration of the collected air gradually decreased as the number of degassing sessions progressed during the experiment. Of course, as the number of degassing sessions increases, the amount of dissolved gas in water decreases, so it can be considered that the oxygen concentration decreases accordingly. However, even if the pressure inside the vessel decreases, the saturated vapor pressure does not change, so the volume fraction or molarity of the vapor will increase. Therefore, even if the composition ratio of the dry air is maintained, the expected oxygen concentration inside the bubble changes (decreases) according to the level of decompression.

Now, let us calculate the composition in the bubble gas as the pressure decreases, assuming that the composition of dry air remains constant. It is assumed that the temperature of the water remains constant of 25°C, and 1 L of water is considered.

1. First, find the vapor fraction inside the bubble.

$$y_{vap} = \frac{p_{vap}}{p_{mix}} = \frac{3.17\,kPa}{p_{mix}}$$

2. Calculate the mass (or number of moles) of oxygen and nitrogen using Henry's law.

$$m_{O_2} = H_{O_2} p_{O_2} = 1.28 \times 10^{-3}\,mole \cdot L^{-1} \cdot atm^{-1} \times p_{mix}\,atm \times \left(1 - y_{vap}\right) \times M_{w,O_2}$$

$$m_{N_2} = H_{N_2} p_{N_2} = 6.48 \times 10^{-3}\,mole \cdot L^{-1} \cdot atm^{-1} \times p_{mix}\,atm \times \left(1 - y_{vap}\right) \times M_{w,N_2}$$

3. Find the volume occupied by the component gas at the pressure p_{mix}.

$$V_{O_2} = \frac{m_{O_2} R_{O_2} T}{p} = \frac{m_{O_2}\,kg \times 0.2598kJ\,/\,kg \cdot K \times 298.15K}{p_{mix}\,kPa}$$

$$V_{N_2} = \frac{m_{N_2} R_{N_2} T}{p} = \frac{m_{N_2}\,kg \times 0.2968kJ\,/\,kg \cdot K \times 298.15K}{p_{mix}\,kPa}$$

4. Calculate the volume of vapor using the known mole fraction of vapor y_{vap}.

$$y_{vap} = \frac{V_{vap}}{V_{O_2} + V_{N_2} + V_{vap}} \quad \text{or} \quad V_{vap} = \frac{y_{vap}\left(V_{O_2} + V_{N_2}\right)}{1 - y_{vap}}$$

The expected values of each component in the bubble calculated by the preceding method are shown in Table 2.B and Figure 2.E. (This represents the fraction of gas in bubbles generated when the pressure is decreased slowly enough to maintain equilibrium condition. The water is assumed to be 25°C saturated solution in equilibrium with the atmosphere that has an idealized dry air composition of 20.9% oxygen and 79.1% nitrogen, respectively.)

Figure 2.E shows that water at 25°C, which was in equilibrium with the atmosphere with an idealized dry air composition of 20.9% oxygen and 79.1% nitrogen, is expected to be in phase equilibrium while the pressure inside the container is reduced from 1 atm to 0.0313 atm. This is the result of calculating the change in composition inside the bubble. As can be seen from the figure, the expected concentration of oxygen from dissolved gas under atmospheric pressure is about 33%, which is much higher than atmospheric pressure. However, as the pressure inside the container decreases, the proportion of vapor within the bubble gradually increases, and the concentration or mole fraction of oxygen and nitrogen decreases.

TABLE 2.B
The Equilibrium Composition of Bubble Gas at Different Levels of Vacuum

p_{mix} (atm)	m_{O_2} (mg/L)	m_{N_2} (mg/L)	y_{O_2}	y_{N_2}	y_{vap}
1	8.29	13.90	0.3322	0.6364	0.0313
0.9	7.44	12.47	0.3311	0.6341	0.0348
0.8	6.58	11.03	0.3296	0.6313	0.0391
0.7	5.72	9.60	0.3276	0.6276	0.0447
0.6	4.87	8.16	0.3251	0.6227	0.0522
0.5	4.01	6.73	0.3216	0.6159	0.0626
0.4	3.16	5.29	0.3162	0.6056	0.0782
0.3	2.30	3.86	0.3072	0.5884	0.1043
0.2	1.44	2.42	0.2894	0.5542	0.1565
0.1	0.59	0.99	0.2357	0.4514	0.3129
0.05	0.16	0.27	0.1283	0.2458	0.6259
0.0313	5.83E-05	9.78E-05	7.46E-05	1.43E-04	0.9998

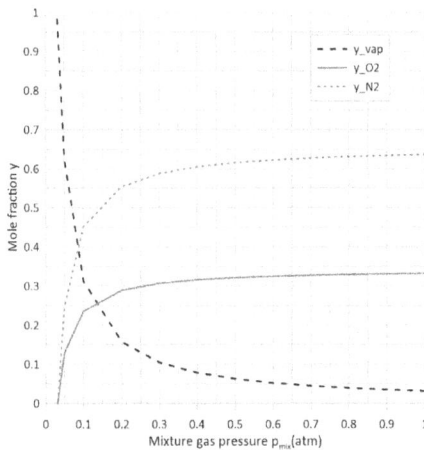

FIGURE 2.E Component mole fractions in the solubility-driven bubble in water (Water temperature is 25°C and the dry air composition is assumed to be composed of 20.9% oxygen and 79.1% nitrogen.)

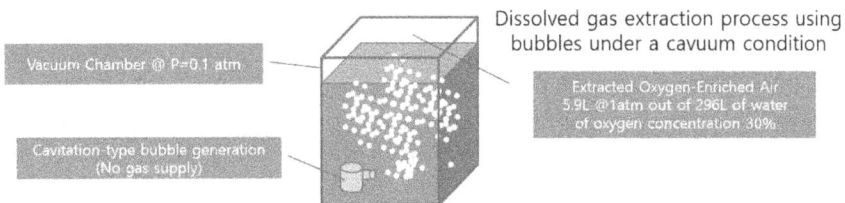

FIGURE 2.5 An illustrative figure that describes the collection of oxygen-enriched-air through vacuum bubbling [19].

An experimental study was conducted to determine whether air with a high oxygen content could be obtained by stripping dissolved gases from water [21]. As shown in Figure 2.5, when the inside of the vessel is depressurized to 0.1 atm and the venturi nozzle flow is created through a submersible pump, bubbles are generated due to the reduced pressure at the nozzle throat. This bubble is not a gas supplied from an external source but is the outgassing of the dissolved gas in the liquid. The amount of gas extracted from 296 L of water at room temperature was 5.9 L under atmospheric pressure, and the average oxygen concentration was 30%.

A specially designed experimental setup (Figures 2.6 and 2.7) was used for this experiment. The pressure vessel, which can withstand a vacuum pressure of 1 kPa, is made of stainless steel with a square cross section and measures 0.65 m (L) × 0.65 m (W) × 1 m (H). It is equipped with two large visualization windows to monitor the situation inside, and a bubbler, which is an assembly of a submersible pump and a venturi nozzle, is installed inside the vessel. A gas collector hood is installed in the middle of the vessel to collect and analyze degassed gas, and a scale is attached on one slanted side surface of the gas collector hood so that the amount of gas inside can be estimated through reading of the water level inside the collector. To facilitate this, 3D CAD tools and ground truth calibration curves are created and used (Figure 2.8). In addition, a bubbler and an air supply line are installed inside the container to bubble with external air. This is used to restore the initial conditions of equilibrium with the atmosphere during repeated experiments. The pressure inside the vessel is reduced by a vacuum generator (ejector: E-Wha Techno Ltd., FOCUS EV-15HS; Republic of Korea) using compressed air supplied by a 5 hP compressor, and the pressure level in the vessel obtained with this system is about 10 kPa (0.1 atm). Three temperature sensors (K-type thermocouples), a pressure gauge (ULFA Technology, SDT

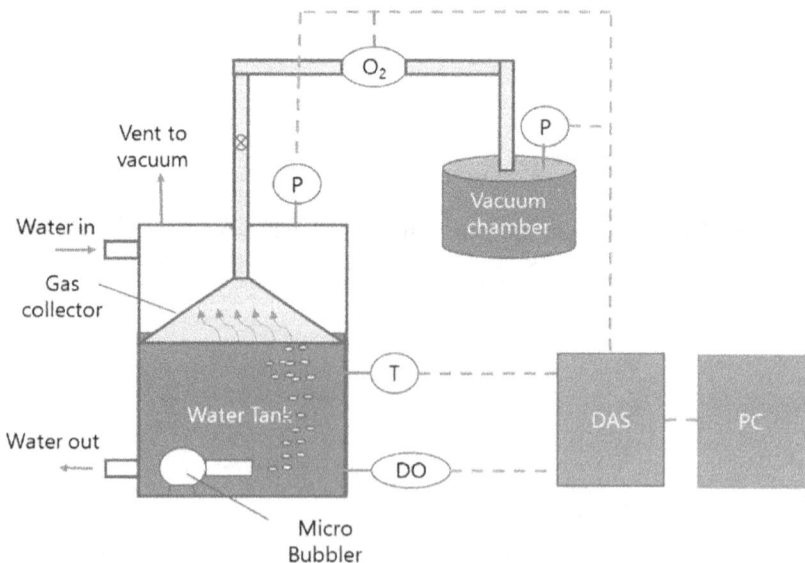

FIGURE 2.6 A measurement system of extracted gas from water [참고문헌].

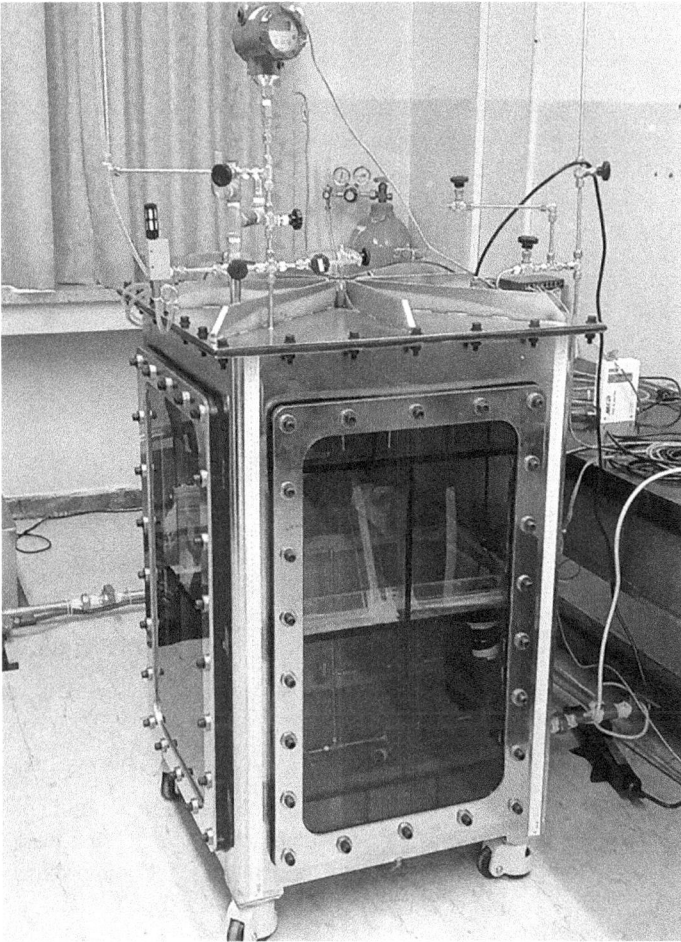

FIGURE 2.7   Water vessel with two large viewing windows for vacuum bubbling experiments [참고문헌].

Series B760H, ±760 mmHg, res.1 mmHg, Republic of Korea), and a dissolved oxygen sensor (Mettler-Toledo InPro 6950i Oxygen Measurement Sensor with M-300 Multi-parameter Transmitter, 6-50,000 ppm, ±5 ppm; Switzerland) were prepared for the measurement system to monitor the condition. An oxygen concentration meter (Alpha-Omega Oxy-Sen Oxygen Monitor, 0-100% oxygen, ±1% of full scale; USA) to measure the oxygen concentration of the captured air was used in connection with a separate vacuum tank. The measured temperature, pressure, dissolved oxygen, and captured air concentrations are stored in a computer through a data acquisition device (Yokogawa DA-100, 30 Ch.; Japan). The submersible pump (Daehwa Electric, Model DPW69-12, 12V DC, 69 lpm; Busan, Republic of Korea) is operated with DC power, and the input power was fixed at about 20 W (nominal) for this experiment. It is revealed through experiments that this selected setting of input power could guarantee the sealing of the pump from overheating of the motor. The experiment

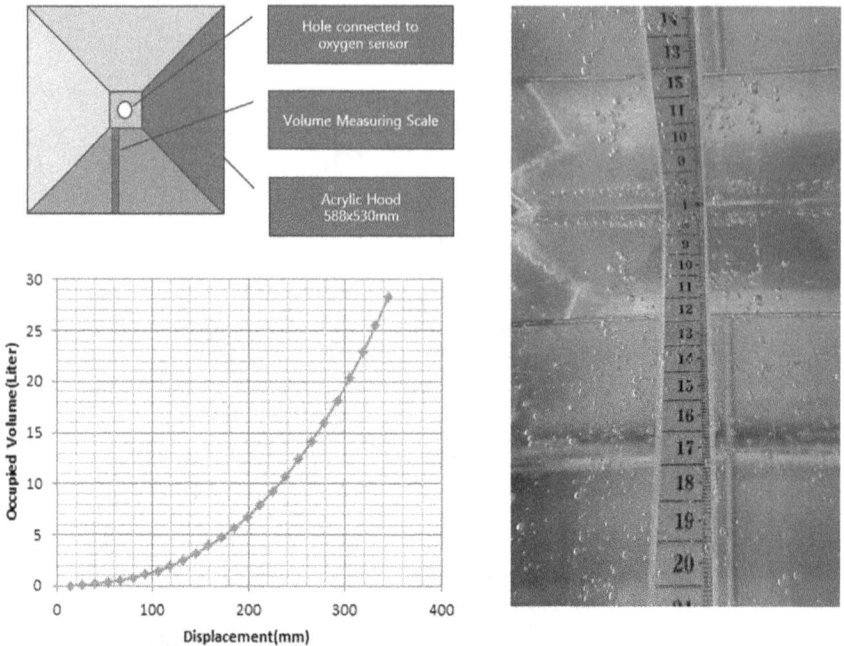

FIGURE 2.8   Extract gas collector hood with a view of measurement scale and a calibration curve.

proceeds by first lowering the pressure inside the container to 0.1 atm and operating the bubbler to generate bubbles. After collecting the gas extracted from the water under the hood of the gas collector to a pre-determined level of approximately 20 to 25 cm based on the scale, bubbling is stopped, and the vacuum is released to recover the pressure to atmospheric pressure. Under the recovered state, the concentration of dissolved oxygen, the amount of captured gas, and the oxygen concentration of captured gas are measured in order.

The testing procedure for degassing and measurement is illustrated in Figure 2.9. Once the sensors for oxygen and dissolved oxygen measurements are calibrated at room condition (0), fresh water is filled in the tank up to the desired level (1), and the initial dissolved oxygen concentration (IDOC) is measured (2). The tank is depressurized down to 0.1 atm with the bubbler being operated. Vapor bubbles are then captured under the gas collector hood (3). Once the inside-the-hood water level reaches the maximum value, the bubbler operation is stopped, and the vent is opened to recover the pressure level back to the atmospheric level (4). Read the inside-the-hood water level and measure the oxygen concentration of the collected gas (5).

Figure 2.10 is a scene where fine bubbles are generated from the bubbler in the initial state of bubbling. In the early stages of bubble formation, a large number of small-sized bubbles are generated like this, making it look like the Milky Way.

In order to differentiate the effect of vacuuming and bubbling, the tests are conducted in two steps, that is, (1) Case 1, depressurization only (0.1 atm or 10 kPa), and Case 2, vacuum bubbling (depressurization combined with bubble generation).

(0) Ullage O_2 and DO sensor calibration.

(1) Prepare fresh water in the tank.

(2) Measure initial D.O. concentration in water.

(3) Depressurization with micro-bubble generation and collection of gas

(4) Open vent port to recover the atmospheric pressure in the ullage

(5) Measure the oxygen concentration and the volume of the collected gas

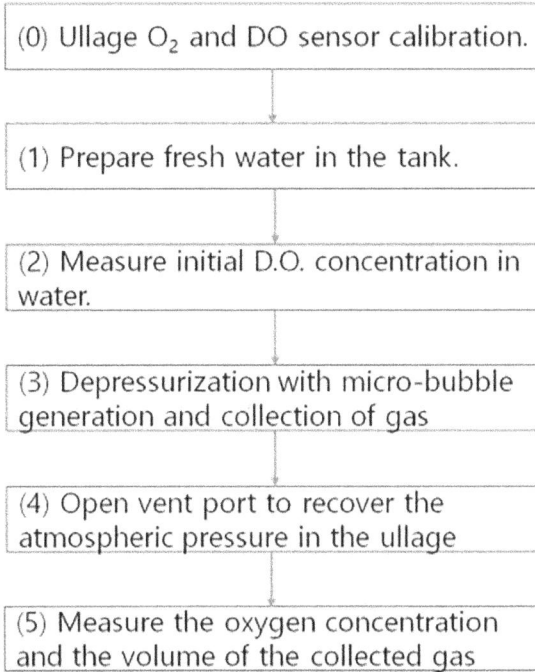

FIGURE 2.9 A testing procedure for degassing and measurements.

FIGURE 2.10 Bubbling view through a venturi bubbler under a vacuum condition ($p = 10$ kPa).

In Case 1, 0.199 L of gas was collected from 296 L of water when the pressure inside the container was reduced to 10 kPa and maintained for 1 hr. Based on an ideal atmosphere of 20.9% oxygen and 79.1% nitrogen in 296 L of water, the expected total dissolved gas volume is 5.47 L when calculated according to the calculation method that follows, and the captured volume is 0.067% (= 0.199 / 296 × 100) of the water volume, a very limited amount corresponding to 3.6% (= 0.199 / 5.47 × 100) of the expected total dissolved gas. The calculation process is as follows.
The mass of each component gas dissolved in the liquid is:

$$m_{O_2} = G_{O_2}V_w = 8.32\,mg\,/\,L \times 296L = 2462\,mg = 2.462\,g$$

$$m_{N_2} = G_{N_2}V_w = 13.89\,mg\,/\,L \times 296\,L = 4111mg = 4.111g.$$

The volume that may be occupied by these gases under an atmospheric pressure becomes:

$$V_{O_2} = \frac{m_{O_2}R_{O_2}T}{P} = \frac{2.462 \times 10^{-3}\,kg \times 0.2598\,kJ\,/\,kg \cdot K \times 298K}{101.3\,kPa}$$

$$= 0.00188\,m^3 = 1.88\,liter$$

$$V_{N_2} = \frac{m_{N_2}R_{N_2}T}{P} = \frac{4.111 \times 10^{-3}\,kg \times 0.2968\,kJ\,/\,kg \cdot K \times 298K}{101.3\,kPa}$$

$$= 0.00359\,m^3 = 3.59\,liter,$$

These results show that dissolved gas extraction in environments without separate means of mixing or surface expansion can be a very slow process.
In Case 2, vacuum bubbling generates bubbles under a reduced pressure of 10 kPa, and in this experiment, degassing sessions are repeated to extract the maximum amount of dissolved gas possible. The total volume of gas captured from the 296 L of test water over five repeated degassing sessions was 5.9 L. Table 2.3 presents the important performance variables during the five degassing sessions, that is, the volume of the entrained gas, the oxygen concentration of the entrained gas, and the dissolved oxygen concentration of the water after degassing. The time required to fill the hoods varied from less than an hour in earlier sessions to over two hours in later sessions. According to the experimental results, the volume average oxygen concentration of the gas collected from 296 L of water was 29.9%, and the dissolved oxygen concentration after degassing was 5.5% O_2 -atm. The oxygen concentration of the captured gas peaked at 31.9% in the second session and tended to gradually decrease in subsequent rounds. This appears to be because the partial pressure of the component gases decreased as degassing progressed. Figure 2.11 shows the measurement results of the oxygen concentration in the extracted gas measured after each degassing session. Measurement of the oxygen concentration of the extracted gas is conducted in the middle of a separate flow path from the collector hood at atmospheric pressure to a separate vacuum tank. As shown in the figure, the oxygen

TABLE 2.3

Degassed Air Measurement Results with Vacuum Bubbling under 0.1 atm (10 kPa)

Test Session	#1	#2	#3	#4	#5
Gas volume (liter)	1.172	1.114	1.405	1.114	1.089
Oxygen concentration (%)	31.0	31.9	30.5	29.3	27.7
DO concentration (%)	15.2	10.8	9.5	7.3	5.5

Source: [21].

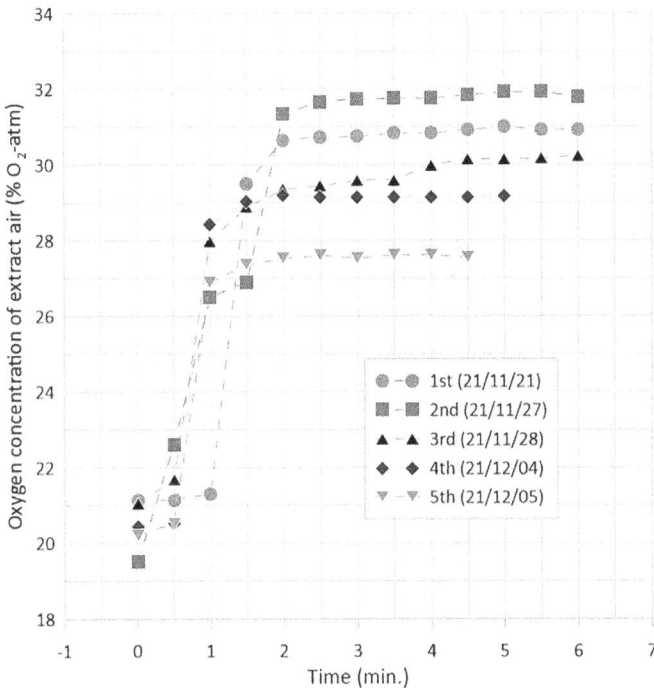

FIGURE 2.11 Measured oxygen concentration of extracted degassed air [21].

concentration gradually increased from the initial atmospheric value as the collected gas passed through the oxygen sensor and was found to maintain a stable value after about 2 min, reaching a maximum value of 31.9% of the measured concentration (session #2), while the minimum value was 27.7% (session #5), and the volume average oxygen concentration was 30.1%. This experimental data is quite close to the expected oxygen concentration of 33.1% as a result of the previous discussion regarding the characteristics of the degassed air from water. As stated in [21], the water temperature at the time of the experiment was 17°C, which is 8°C lower than the initially assumed 25°C, and in that paper, through temperature correction, the total amount of gas dissolved in water was re-estimated to be 6.30 L.

Summarizing the degassed air analysis from water through vacuum bubbling, the experiment showed that the degassed air had an oxygen concentration 50% higher than that of the atmosphere, and that a significant part of the dissolved gas could be degassed through vacuum bubbling. However, the maximum amount of gas that can be trapped in water is about 2% of the liquid volume at atmospheric pressure. This information can serve as a reference for researchers trying to directly apply artificial gills when estimating the amount of water that must be processed to provide the amount of oxygen required for human respiration.

2.1.4 Solubility Data for Interesting Fluids Other Than Water

One of the applications where solubility and outgassing in fluids other than water are important is jet fuel. In general, it is known that the gas solubility of petroleum-based liquids is higher than that of water [22], and proper handling of dissolved gas has a very important meaning. One example is the story of aircraft safety. Why am I suddenly talking about airplanes? That is a good question. As you know, aircraft are designed to fly at very high altitudes to travel quickly with less drag. In the case of the international flights, it flies at an altitude of about 10,000 m, and the atmospheric pressure at this altitude is about 1/4 of the surface level [23], which means that about 3/4 of the dissolved oxygen that was in equilibrium at the surface of the ground at 1 atmosphere becomes ready to escape from the fuel. Because it can escape with a high oxygen concentration, gas with a high oxygen concentration released in this way can be viewed as a potential risk factor for the safe flight of an airplane in general.

According to Frank and Drikakis [24], malfunction of the equipment of an aircraft can result in sparks being generated within the fuel tank, and the composition of gases inside the ullage, the space in a fuel tank over the fuel, determines the probability of ignition in response to such sparks [25]. After the Trans World Airlines Flight 800 (TWA 800) disaster, caused by the ignition of flammable vapors in the fuel tank [26], awareness of the importance of fuel tank safety in aircraft is raised, and research was initiated to find measures to keep the fuel tanks in an inert state. One effective way to reduce the risk of combustion in a fuel tank is known to reduce the oxygen concentration in the ullage to a level under which combustion does not proceed [27]. In this regard, the Federal Aviation Administration (FAA) considers a fuel tank of a commercial aircraft inert if the oxygen percentage is below 12% by volume [28]. Currently, nitrogen-enriched air (NEA) is pumped into the fuel to reduce the oxygen concentration in the ullage and render it inert [29].

So what is the solubility of atmospheric gases in fuels? According to the ASTM standard [22], the gas solubility of petroleum liquids with density $d = 0.85$ is presented as the value of the Ostwald coefficient. The Ostwald coefficient is defined as the solubility of a gas expressed as the volume of dissolved gas per volume of liquid when the gas and liquid are in equilibrium at a specified gas partial pressure and a specified temperature. Based on the solubility data of oxygen in petroleum liquids mentioned earlier in the standard, when the Oswald coefficient of 0.174 under 25°C atmospheric pressure is converted, it is about 50 mg/L, which is about six times greater than the solubility of oxygen in water of 8.3 mg/L under the same condition. Solubility conversion from Ostwald coefficient to mg/L can be done according to the relationship given in the standard. In order to respond to the explosion safety of aircraft fuel containing such a large amount of oxygen, a method of maintaining a low oxygen concentration by washing the ullage with nitrogen-enriched air is currently being adopted. If the burden on the combustion performance or other side effects of the fuel is not great, how about applying pre-deaeration to the fuel? As introduced earlier, vacuum bubbling has a simple structure and can deaerate a large amount of dissolved gas in a short time. Of course, there are probably countless things to consider in this part, such as performance, stability, and operability. However, it is also believed that the role of a preliminary measure to secure the safety of the astronomically increased number of air passengers around the world is already in place. It is always interesting to envision how the future of mankind will unfold. Now the story goes back to the water, to the most serious subject of this book: the vapor bubble.

2.2   VAPOR PRESSURE AND VAPOR BUBBLE GENERATION

The second stage of vacuum bubbling involves the generation of vapor bubbles. This is completely different from the first stage of vacuum bubbling, in which the supersaturated solute is separated from the liquid in the form of bubbles due to the limit of solubility. In other words, when a phase change occurs under a pressure lower than the saturated vapor pressure of the liquid, the main component of the bubbles becomes vapor, and at this time, the concentration of dissolved oxygen present in the space where the bubbles are formed can no longer be expressed in terms of solubility. Instead, the volume fraction or pressure fraction applicable to the mixed gas should be used. The concentration of dissolved gas inside the vapor bubbles formed in this way is usually very low, practically close to zero, and therefore, degassing by vapor bubbles should be understood as mass transfer by stripping or diffusion, which removes components by concentration difference. Additionally, vapor bubbles created by vacuum bubbling are cavitation bubbles in nature, but if the pressure around the bubbles is sufficiently low, there is little concern about the undesirable behavior of ordinary cavitation bubbles. Figure 2.12 is a phase change diagram for water. The saturated vapor pressure of a substance is a function of temperature alone, and if the pressure is higher than the saturated vapor pressure, it is in a liquid phase, and if the pressure is lower than it, it is in a vapor phase. Let us begin our discussion of the generation and retention of vapor bubbles, keeping this diagram in mind.

In a usual practice, in order to obtain vapor bubbles, heat is added to the system so that the temperature of the system increases to the saturation temperature (boiling temperature) while maintaining the initial pressure level, then bubbles are initiated typically at locations of superheated surface. In this process, heat energy is required for raising temperature and for phase change, known as evaporation. In the meantime, to achieve the same goal of vapor bubble generation, a phase-change phenomenon, we can lower down the pressure below the saturated vapor pressure, maintaining the initial temperature (in other words, without any heating). In this case, vapor bubbles are created at a pressure lower than the saturated vapor pressure. The vapor bubble generated in this way is called cavitation bubble. Typically, the cavitation bubbles are

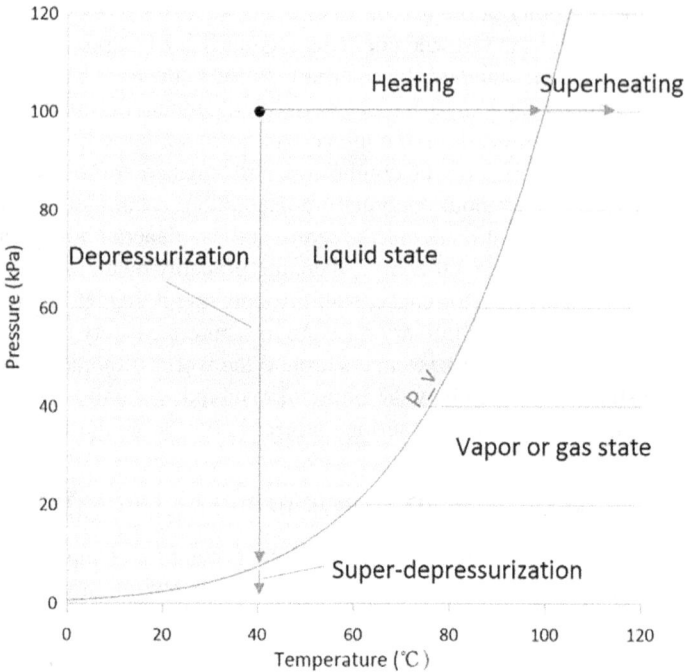

FIGURE 2.12   Two approaches to arrive at phase-change condition: heating vs. depressur-ization (or vacuuming).

generated at the suction side of the impeller blades, as in the case of the propeller of ships. Heating and superheating, and the generation of vapor through heating, have been studied extensively so far for the applications in thermal engineering fields, such as boiler for steam generation. Now, let us take a look at another possibility of generating vapor bubbles by depressurization. Similar to the case of heating and superheating, vapor bubbles can also be generated through any kinds of energy input, such as mechanical work, through which phase change may occur, resulting in the generation of vapor bubbles.

2.2.1   Bulk vs. Local Conditions

The phase diagram shown in Figure 2.12 assumes a uniform condition; however, the body of water is hardly homogeneous in terms of temperature, and the vapor bub-bles are described quite often to be generated on the superheated surface on which the local thermodynamic condition is ready for making phase change from water to vapor. Until the temperature of the bulk body of water reaches the saturation temper-ature, the vapor bubble cannot grow or maintain the initial bubble state, because the condition outside the bubble is not yet thermally matured. Bubbles finally collapse or shrink after they depart from the superheated surface, and the trace of bubbles starts navigating in the bulk water. This is the situation you have when you boil water in the pot for cooking. If water fully reaches the saturation temperature in a homogeneous manner, then the vapor bubble may be able to maintain its form of vapor without shrinking and/or even may grow. This will define the condition of massive bubbling. The boiling (vaporization through heating) process is so widely used even in everyday life that it is recognized that you heat and water boils. An important point is that in order to make water change phase, the location of superheated region (local) should exist in the system, and depending on its degree of superheating (thermal energy input), the rate will be varied. In this context, we can understand the vapor bubble generation process (other than the surface evaporation) through heating is composed of two stages, that is, (1) heating and (2) superheating. Also, it needs to be emphasized that there are, even in the case of heating, factors that affect the bubbling conditions, such as the depth of the heating surface (hydrostatic effect), water temperature, vessel pressure, and heat input or superheat exist. Just as the two-step approach worked well for heating and superheating, it can also be useful for vacuum bubbling. The first step is to depressurize the entire space within the vessel close to the saturated vapor pressure, and the second step is to create local pressures lower than the saturated vapor pressure, which directly correspond to heating and superheating in the heating counterparts. Conceptually, these correspond exactly to the two phases of heating and superheating in the heating process.

Now, let us start with the second phase of the discussion of vacuum bubbling as is introduced in Figure 2.13. The first depressurization is a step of lowering the pressure inside the vessel. The low pressure inside the vessel creates an environment in which the supersaturated solute can easily escape in the form of bubbles. In addition to this, lower pressures can be created locally within the solution, for example, using a venturi nozzle coupled with a small submersible pump. When the pressure drops below the saturated vapor pressure, you may think that the liquid turns into vapor and vaporizes, but this is not always the case.

Figure 2.14 is an example of a typical phase-change diagram [1]. In the region where the decompressed pressure is slightly higher than the saturated vapor pressure

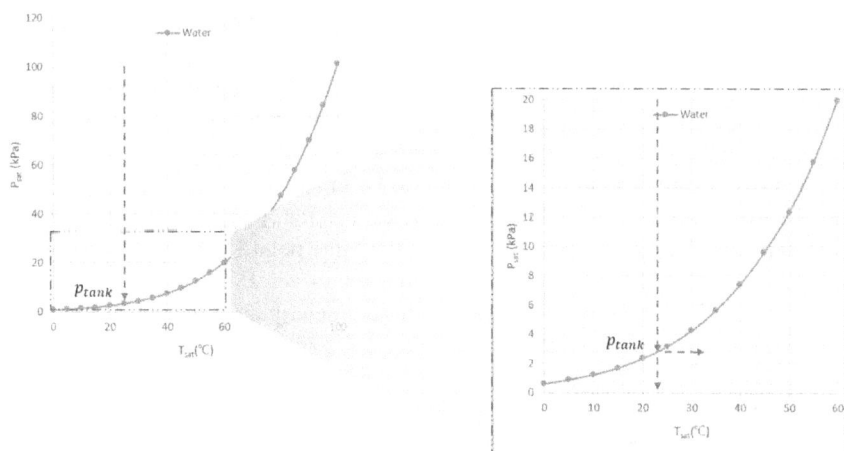

FIGURE 2.13 A two-step approach to generate vapor bubbles.

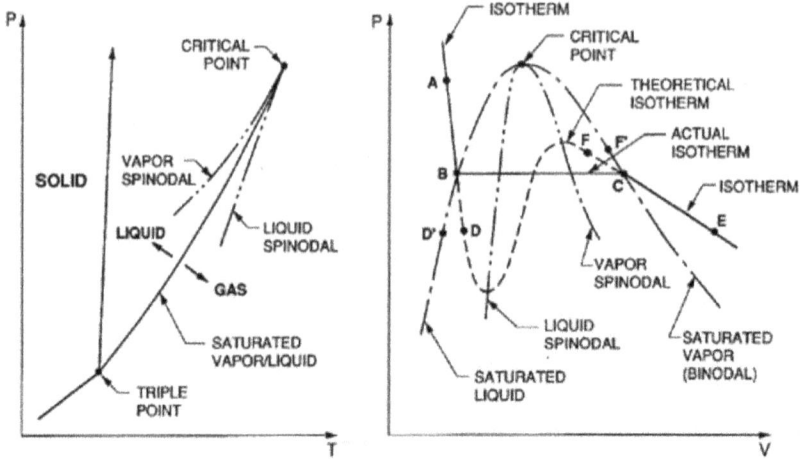

FIGURE 2.14 Typical phase diagrams that explain the existence of metastable region and the spinodal lines [1].

(positive pressure: +), the existing vapor (bubble) maintains the vapor state, and conversely, in the region slightly lower (negative pressure: −), the metastable region maintains the existing liquid state, respectively (Figure 2.14(a)). (In this discussion, since we are talking about the action of depressurization, we intentionally do not mention the effect of heating or cooling.) In thermodynamics, the limit of local stability for small fluctuations is defined by the spinodal curve, beyond which extremely small fluctuations in composition and density cause phase separation via spinodal decomposition, and between the saturation line and the spinodal line, the solution is at least metastable to the fluctuations [1]. In the metastable region, if the energy required for phase change is not sufficient, the liquid state remains intact, and at this time, it is known to maintain a state that is no longer amplified even when perturbations such as pressure changes occur. For reference, in the phase diagram of Figure 2.14(b), if the pure liquid in state A is depressurized at constant temperature and the pressure is reduced below point B (saturated vapor pressure) without significant nucleation sites, decompression can continue. A liquid that falls along the theoretical isotherm to point D is said to be in tension, and the pressure difference between B and D is the magnitude of the tension, while if the temperature is increased and proceeds along the isobars at the same point D' to reach point D, it is called superheat, and the temperature difference between D and D' is called degree of superheat [1]. Thermodynamically, the phase is determined by whether the energy (Δh_{fg}) required for the phase change of a material is sufficiently supplied or not, and the amount of energy transfer required for the phase change in this way is the extent of tension and duration, according to Brennen [1]. In the end, it can be seen that generating vapor bubbles according to the desired conditions and securing stable conditions for generating a large amount seems to be the most important technical issue in vacuum bubbling.

2.2.2 AN OPERATION MODEL FOR VACUUM BUBBLING

As a way to understand the vapor bubble generation conditions in more detail, an operation model for vacuum bubbling, more precisely, vapor bubble generation, was presented for the experimental device model through the venturi nozzle bubbler in the pressure-reducing tank. Figure 2.15 is a virtual vacuum bubbling device for theoretical review of vacuum bubble generation, and its approximate shape and function are as follows:

The inside of the container is filled with water, and a vent for depressurization and gas discharge is connected to a vacuum device at the top of the container. A bubbler nozzle is installed horizontally at a depth Δh in the water, and this nozzle is connected to a submersible pump operated by DC power. For the sake of discussion, for convenience, the room condition is denoted by subscript 0, the empty space inside the container is denoted by subscript 1, the inlet of the nozzle is denoted by subscript 2, the nozzle throat is denoted by subscript 3, and the downstream area of the nozzle is denoted by subscript 4, respectively. Room temperature (T_0) and water temperature (T_w) may need to be considered independently, but only the water temperature will be considered in this discussion. The power supply of the submersible pump that drives the bubbler is noted to be DC just because of the availability for small pumps.

To discuss the conditions for the second-stage bubbling, the thermodynamic state during the vacuum bubbling process is displayed on the phase diagram shown in Figure 2.16. Of course, this model of using a nozzle to generate bubbles is just one of possible approaches, but observations on this model are expected to provide a more reasonable understanding of the vapor bubble generation model. In addition, among various physical variables, variables corresponding to the available experimental conditions were fixed. That is, the water temperature $T_w = 25$, the nozzle depth $\Delta h = 0.8$ m, the tank ullage pressure $p_1 = 5\,kPa$, and a fixed flow rate through the nozzle defined by a constant DC power of 23.4 W (7.8 V and 3.0 A) define the baseline condition. This model is a basic performance model of a venturi bubbler

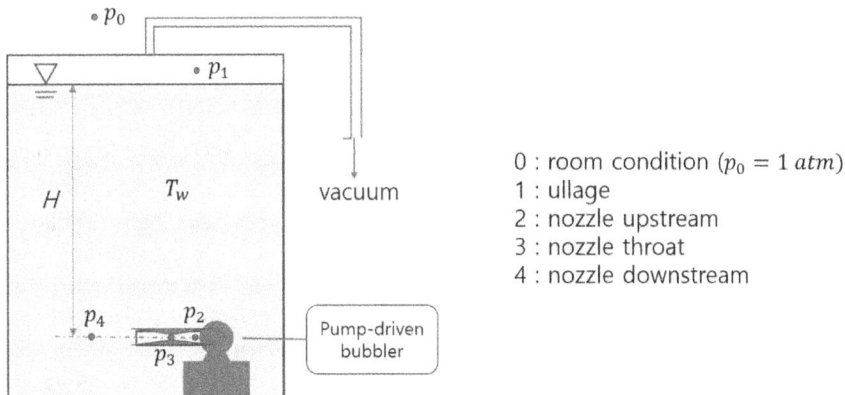

0 : room condition ($p_0 = 1\ atm$)
1 : ullage
2 : nozzle upstream
3 : nozzle throat
4 : nozzle downstream

FIGURE 2.15   A physical model of vacuum bubbling process.

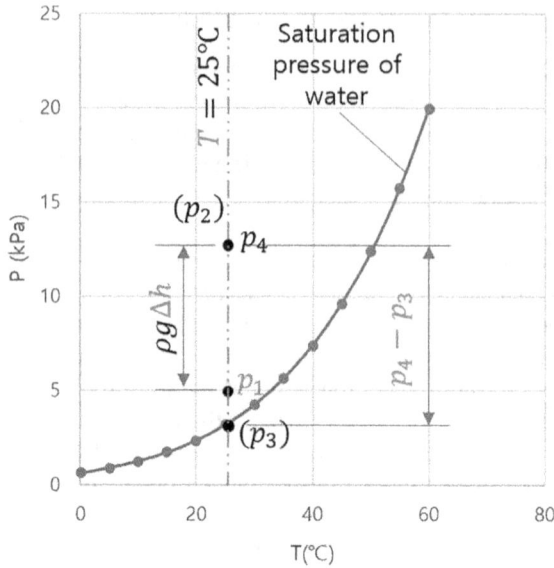

FIGURE 2.16 A baseline operating condition model for vacuum bubbling experiment ($\Delta h = 0.8\,m$).

to describe the generation and survival conditions of vapor bubbles. One of the two steps for vacuum bubbling is to reduce the pressure inside the vessel by the vacuum pump to p_1, and the other is to reduce the local pressure by the flow induced by the pump. When the flow is formed from the pump, the pressure at each part of the nozzle is represented by the inlet p_2, the nozzle throat p_3, and the nozzle downstream p_4, as shown in Figure 2.16. By the way, the pressure downstream of the nozzle p_4 depends on the depth of the nozzle Δh, such that $p_4 = p_1 + \rho g \Delta h$ due to the effect of the hydrostatic pressure at that depth. From this, we can derive four important performance design variables that can directly affect the generation of bubbles. These **four performance variables are decompression level p_1, nozzle depth Δh, water temperature T_w, and nozzle performance $p_4 - p_3$**, respectively. Here, the nozzle inlet pressure p_2 and the nozzle throat pressure p_3 are values that must be theoretically or experimentally determined.

Among the four performance variables, the nozzle throat pressure p_3 is a variable directly related to the generation of bubbles. When the pressure at the nozzle throat becomes lower than the saturated vapor pressure, the difference between the saturation pressure and the nozzle throat pressure $p_{sat} - p_3$ becomes the magnitude of tension, which seems to determine the bubble generation pattern. According to the observation results through the experiment, it was observed with the naked eye that the higher the magnitude of tension, the larger the amount of bubbles and the larger the bubble size. In conclusion, $p_{sat} - p_3$ is the most important variable determining vapor bubble generation. Tension is a combined result from (1) the level of decompression p_1, (2) the depth of the nozzle Δh, (3) the flow condition of the nozzle $p_4 - p_3$, and (4) the temperature of the water T_w. Among them, the flow condition of the nozzle is determined by the flow rate according to the nozzle shape and input power. When a bubble is formed near the nozzle throat, the cross-sectional area of the passage downstream of the nozzle throat increases, so according to Bernoulli's theorem, the flow velocity decreases and the pressure increases, increasing from p_3 to p_4. At this time, if p_4 is much higher than the saturated vapor pressure and leaves the metastable region, the vapor bubble generated at the nozzle throat disappears through a rapid decrease process. Therefore, in order to maintain the generated vapor bubble, the nozzle downstream condition p_4 that is determined by the tank internal pressure p_1 and the depth of the nozzle Δh is particularly important, that is, $p_4 = p_1 + \rho g \Delta h$. After all, the difference between p_4 and the saturated vapor pressure p_{sat} is critical to the survival of vapor bubbles. If the vapor bubble generated in the nozzle shrinks or disappears at the nozzle exit, the mass transfer role of the vapor bubble will be extremely limited.

Now, let us discuss the operating conditions for the baseline operation shown in Figure 2.16. Three of the four variables are values that can be simply set, but among them, $p_4 - p_3$ is a variable dependent on the flow rate through the venturi nozzle, and there are limitations in directly checking them. Nonetheless, for the discussion of the parameters, for the venturi nozzle shape shown in Figure 2.17 with a nozzle neck diameter of 8 mm, given a pump input of 23.4 W, the flow rate through the nozzle and the pressure difference between the nozzle outlet and the nozzle throat were roughly estimated using energy equation [30] to be 42 lpm and $p_4 - p_3 = 9.65$ kPa. For this estimation, an assumed pump efficiency of $e = \dfrac{\dot{W}_{s,out}}{\dot{W}_{s,in}} = 0.5$ and minor loss coefficients for entrance, gradual contraction, and expansion are considered. (Although this estimate is imprecise, it can be safely regarded as an assumption for the sake of qualitative discussion.)

From the preceding discussion, the baseline vacuum bubbling conditions are defined as follows:

- $p_{sat} = 3.17\,kPa$ at $T_w = 25$
- $p_1 = 5\,kPa$
- $p_4 = p_1 + \rho g \Delta h = 5 + 997 \times 9.81 \times 0.8 / 1000 = 12.8\,kPa$
- $p_3 = p_4 - (p_4 - p_3) = 12.8 - 9.65 = 3.15\,kPa$

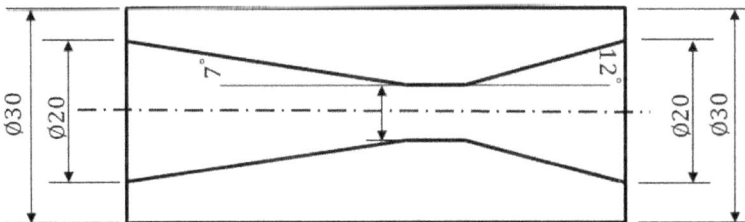

FIGURE 2.17   A venturi-nozzle configuration used for bubble generation experiments (Flow direction is from right to left; not in scale).

In this case, the magnitude of tension $p_{sat} - p_3$ at the nozzle throat, which is the condition for generating vapor bubbles, is $3.17 - 3.15 = 0.02$ kPa, which is very insignificant, but it can be seen that there is room for bubble generation. However, since $p_4 - p_{sat}$, the retention condition of the vapor bubble, is $12.8 - 3.17 = 9.63$ kPa, which is more than three times the saturated vapor pressure of water at 25°C, it can be seen that the vapor bubble exiting the nozzle is difficult to survive anymore. Let us begin our discussion of the four performance parameters described earlier. To this end, the other three variables are reviewed one by one in a fixed state.

2.2.2.1 Change in Water Temperature (T_w)

By raising the operating water temperature, the corresponding saturated vapor pressure increases. An extreme case would be heating without changing pressure, in which water temperature will eventually reach a 100°C. If we lower down the pressure in this condition, bubbling would be massive enough that depressurization itself may be a challenge because the rate of bubble generation may be massive. Another issue in this condition would be the stability of the vaporization process, because depending on the level of vacuuming, the thermodynamic state may reside in a point of unstable region lower than the spinodal line (see Brennen [1]). As the water temperature (T_w) increases, the saturated vapor pressure also increases, resulting in a greater magnitude of tension at the nozzle throat ($p_{sat} - p_3$), and the pressure downstream of the nozzle can become closer to the saturated vapor pressure. This result ultimately helps secure more favorable bubble generation conditions ($p_{sat} - p_3$) and more favorable bubble retention conditions ($p_4 - p_{sat}$).

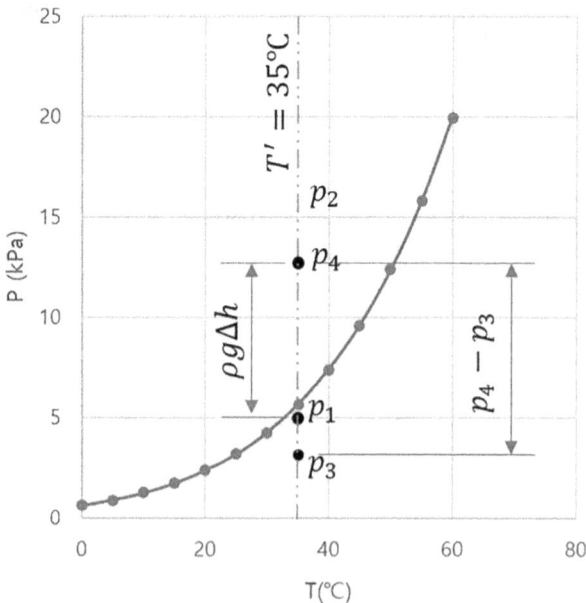

FIGURE 2.18 An operating condition model for a raised temperature ($T'_W = 35$°C).

For example, let us go back to a moderate heating to raise the water temperature from 25°C to 35°C. Maintaining the same conditions other than the temperature of water, the operating conditions are illustrated in Figure 2.18. By having a temperature increase by 10°C from 25°C to 35°C, the operating conditions are formed with the saturated vapor pressure rising to about 2.52 kPa (from 3.17 kPa to 5.63 kPa), the degree of tension at the nozzle throat increases from 0.02 kPa to 2.5 kPa, and $p_4 - p_{sat}$ at downstream of the nozzle reduces from 12.8 − 3.17 = 9.63 kPa to 12.8 − 5.63 = 7.71 kPa. In this case, the vapor bubble generation conditions at the nozzle throat can be expected to directly improve, but the difference between the pressure downstream of the nozzle and the saturated vapor pressure $p_4 - p_{sat} = 7.71 kPa$ is still a significant difference. It can be considered a condition in which the vapor bubbles generated at the nozzle throat easily shrink or disappear.

2.2.2.2   Change in the Nozzle Depth from Water Surface Δh (The Effect of Hydrostatic Pressure)

When water is to be heated and to be boiled (see Figure 2.19), the usual way of heating is the bottom heating, as illustrated (Figure 2.19(a)). The main mechanism of heat transfer in this situation is the combination of conduction, convection, and radiation, all of which require temperature difference or gradient for the transfer of heat. When the temperature of bulk water is still below the saturation temperature, the local difference in temperature provides the condition for convection (free convection), and the energy that is absorbed from the bottom surface tends to diffuse to reach a uniform temperature field. Once the temperature reaches the saturation temperature or higher on the heated surface, water starts changing phase to vapor in the form of bubbles. When the bulk water temperature is not yet heated enough, having a lower temperature than the saturation temperature, the generated bubbles face a kind of adverse condition to maintain its existence upon departure from the bottom surface. That is why the tiny bubbles at the early stage of heating by ovens start being generated and disappear or diminish in size upon rising through the rest of the water body.

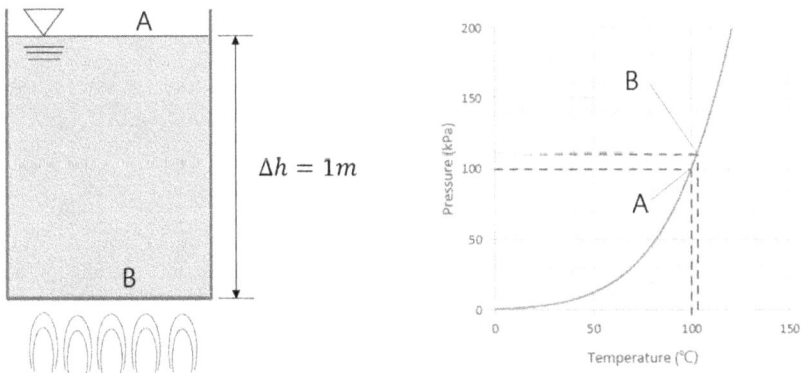

FIGURE 2.19   The effect of hydrostatic pressure on phase-change condition in the case of heating.

In order to look into a little more detail on the behavior of bubble generation, suppose we have a container filled up with water of 1 m depth, as shown in
Figure 2.19(a). On the water surface A that is exposed to atmosphere of 101.3 kPa,
the saturation temperature is 100°C, as you expect, as shown in Figure 2.19(b).
However, on the bottom surface B, the local pressure is no longer the same as 1 atm;
rather, it is higher than 1 atm due to the hydrostatic pressure contribution, that is,
$p_B = p_A + \rho g \Delta h \cong 101.3 + 997 \times 9.81 \times 1/1000 = 111.1 kPa$. The corresponding saturation temperature then would be 102.7°C. This difference may not be considered
appreciable, but it is clear that even though you have reached 100°C, water would not
start boiling at all at that temperature.

However, in the case of vacuum bubbling, the role of hydrostatic pressure seems
no longer trivial, because the pressure level we deal with is the order of kPa rather
than 100 kPa of atmosphere. The saturated vapor pressure of water at 25°C is known
to be 3.1698 kPa, and the hydrostatic pressure contribution with the depth in water is
around 1 kPa per 0.1 m depth ($\rho g \Delta h = 997 kg/m^3 \times 9.81 m/s^2 \times 0.1 m = 0.98 kPa$).
So if you have a bubbler at 1 m depth of water, the contribution of the hydrostatic
pressure itself becomes much larger than the saturated vapor pressure. This vapor
bubble generated at or immediate downstream of the throat are supposed to be collapsed due to the high-pressure condition downstream, p_3 to p_4.

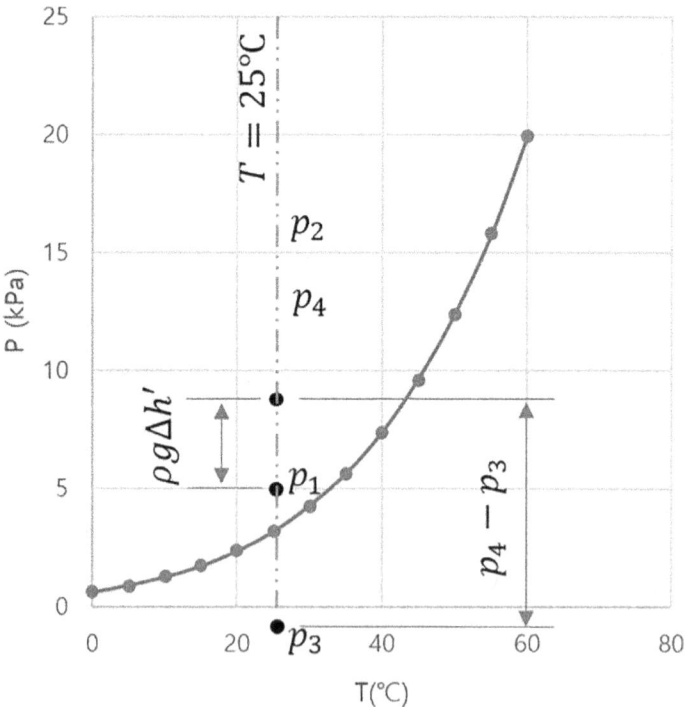

FIGURE 2.20   An operating condition model for a reduced nozzle depth ($\Delta h' = 0.4 m$).

Suppose, for example, that the depth of water Δh of the nozzle is changed from 0.8 m to 0.4 m; the change in $p_{sat} - p_3$ corresponds to the difference in hydrostatic pressure acting on the nozzle. Therefore, the magnitude of tension increases by $\Delta(p_{sat} - p_3) = \rho g \Delta h = 997 \, kg / m^3 \times 9.81 m / s^2 \times 0.4 m = 3.9 \, kPa$. In this case, it is believed that the pressure difference inside the nozzle will not actually appear as much as $p_4 - p_3$, as shown in Figure 2.20, due to the effect of bubbles generated inside the nozzle, but it is believed to affect the magnitude of tension.

In this case, the pressure at the nozzle throat p_3 and downstream of the nozzle p_4 decreases by 3.9 kPa as the nozzle depth changes, so the pressure at the nozzle throat is expected to be $p_3' = 3.15 - 3.90 = -0.75$ kPa, and the pressure at the nozzle downstream is $p_4' = 12.8 - 3.9 = 8.9$ kP, respectively. For the nozzle throat pressure is an estimated value based on the assumption that the nozzle is filled with liquid, it should be considered that the bubbles generated at the nozzle throat might have changed the flow conditions. However, the change of nozzle depth causes the change in the magnitude of tension at the throat theoretically by $3.17 - (-0.75) = 3.92$ kPa, and in the pressure condition at the nozzle downstream to be $p_4 - p_{sat} = 8.9 - 3.17 = 5.73$ kPa.

2.2.2.3 Change in the Ullage Pressure p_1

The global vacuum level p_1 is especially important a parameter that affects the entire mechanism and the performance of vacuum bubbling. If we have a higher pressure for p_1 than 5 kPa in the baseline condition, for example, keeping other conditions the same, then there is zero possibility of getting vapor bubbles, because in that case the minimum pressure at throat p_3 becomes greater than the saturated vapor pressure p_{sat} at that temperature. The generated bubbles in this condition originate from the dissolved gases, which come from the supersaturated portion of the solute gases, but not from water vapor. In this case, bubbles have higher oxygen concentration than that of the atmospheric air due to the higher solubility of oxygen than nitrogen in water. This is interesting and also finds applications for extracting high-oxygen-content air from water. This gas extracts may be utilized as substitute for breathing air when the atmospheric air is not available, such as in underwater breathing. The pressure at the nozzle downstream p_4 will provide different situations, depending on its value, compared to the saturated vapor pressure. If $p_4 > p_{sat}$, the vapor bubble generated at or downstream of the nozzle throat is eventually exposed to the thermodynamic state of liquid, which eventually drives the vapor bubbles to be condensed or collapsed. There is, though, a possibility of bubbles survival due to the lagged phase change or due to the entrained non-condensable gases in the bubble.

Suppose that p_1 is decreased from 5 kPa to 1 kPa; for example, as shown in Figure 2.21, which is lower than p_{sat}, the resulting pressure at nozzle throat where the bubble generation supposedly may occur becomes lower and the magnitude of tension, $p_{sat} - p_3$, may increase at nozzle throat, which is beneficial to the bubbles generation. However, for the nozzle downstream pressure, p_4, still remains higher than p_{sat}, bubbles are supposed to go through an adverse pressure region where vapor condensation or collapse may occur. When the pressure in the tank is reduced from 5 kPa to 1 kPa, the magnitude of tension may increase by 4 kPa, as shown in Figure 2.21. In this case, the estimated pressure at the nozzle throat becomes $p_3' = 8.82 - (p_4 - p_3) = 8.82 - 9.65 = -0.83 \, kPa$, and the pressure at the nozzle

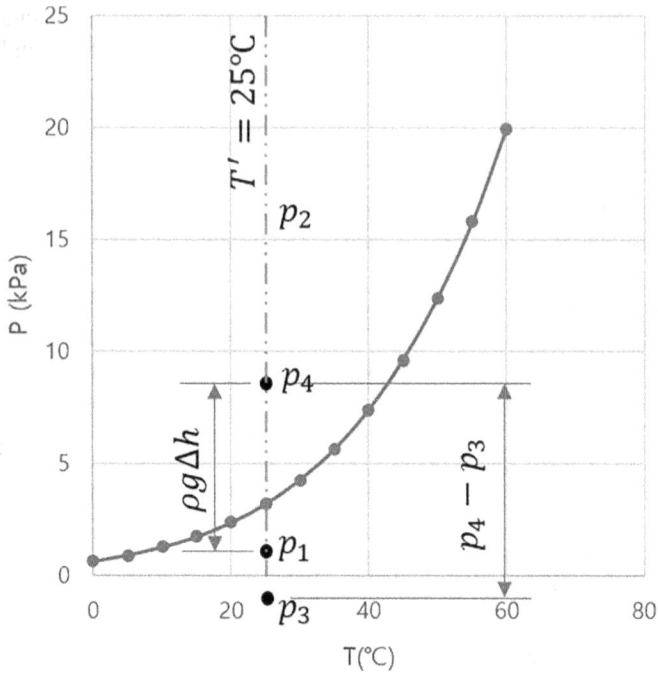

FIGURE 2.21   An operating condition model for a reduced vessel pressure ($p_1' = 1\ kPa$).

downstream is $p_4' = 1 + 997 \times 9.81 \times 0.8 / 1,000 = 8.82$ kPa. The nozzle throat pressure is a value estimated based on the liquid nozzle, and it should be considered that it is different from the actual pressure when bubbles are generated at the nozzle throat and the flow conditions change accordingly. The change in p_1 causes the change in the magnitude of tension at the throat theoretically to $3.17 - (-0.83) = 4.00$ kPa, and in the pressure condition at the nozzle downstream to be $p_4 - p_{sat} = 8.82 - 3.17 = 5.65\ kPa$.

2.2.2.4 Change in the Nozzle Performance $p_4 - p_3$ According to Pump Design and Pump Input Power

If we provide more power to the pump, which may increase the flow rate, resulting in higher pressure difference between the throat and the exit or downstream along the centerline, $\Delta p = p_4 - p_3$, the rate of bubble generation at throat would increase; however, so far as the downstream condition, p_4, remains higher than the saturation pressure, p_{sat}, the effect of increasing power may be limited to the situation that it increase the magnitude of tension at throat and the duration time thereof.

If the input power of the pump is increased, the flow rate and the pressure difference between the nozzle outlet and nozzle throat will increase. If the pressure difference between the nozzle outlet and the nozzle throat increased to 11.8 kPa, for example, as shown in Figure 2.22, ideally, the pressure at the nozzle throat reaches 1 kPa, resulting in the magnitude of tension $p_{sat} - p_3' = 3.17 - 1 = 2.17$ kPa, and the nozzle downstream pressure condition becomes $p_4' - p_{sat} = 12.8 - 3.17 = 9.63$ kPa.

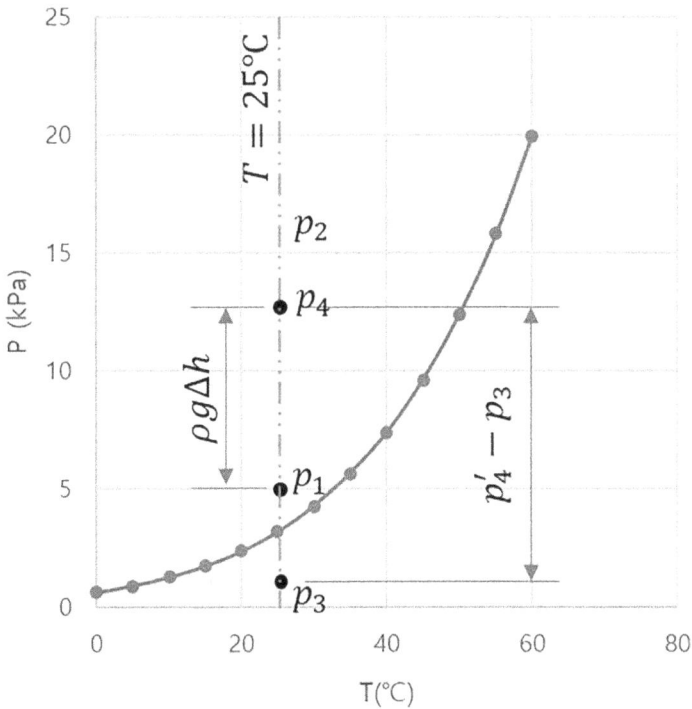

FIGURE 2.22   An operating condition model for an increased flow rate ($p_4 - p_3 = 11.8\,kPa$).

Although not included in the discussion, it should not be overlooked that the performance of the nozzle depends on the shape design of the nozzle prior to the input power. Also, from a practical point of view, if the power of the pump increases and the heat in the motor increases, it may become difficult to maintain perfect sealing due to the contraction–expansion action of the gas filling the empty space inside the motor due to the temperature difference.

From the preceding discussion, important parameters in vacuum bubbling could be identified. The pressure condition downstream of the nozzle, p_4, is especially important in a sense that p_4 is the final highest pressure the vapor bubbles are exposed to. If $p_4 > p_{sat}$, bubbles will collapse and if $p_4 \leq p_{sat}$, bubbles will survive and grow. Once the bubbles survive and navigate through water till they arrive at the water surface, the degassing contribution from these bubbles would be significant because the bubbles may have good enough time, size with a considerable concentration gradient for diffusion mass transfer. In practice, however, the slot of pressure range is quite narrow at low temperature. For example, at 25°C, the saturated vapor pressure of water is known to be 3.17 kPa. In order to have $p_2 \leq 3.17$ kPa, the sum of $p_1 + \rho g \Delta h$ should remain lower than this value, which limits the depth of the bubbler nozzle. In order to remedy this limitation, a moderate heating can be a compromised solution, and other implementation techniques need to be devised. A key message can be said that the pressure downstream of the nozzle, p_4, will determine the immediate destiny of the vapor bubble generated at nozzle. If p_4 can be maintained lower than the local p_{sat}, this will constitute the condition for massive vapor bubble generation. If p_4 is moderately higher than p_{sat}, part of bubbles that survive at the downstream will rise up due to its buoyancy to a lower pressure region (bubbles move upward). In that case, bubbles may survive till they arrive at the surface. Other than those, vapor bubbles generated at throat will face adverse condition to maintain its phase, and it seems a matter of time and extent of survival because of the thermodynamic condition that belongs to a liquid state.

Suggestions for designing a vacuum bubble generating system using the four design factors presented previously are as follows:

1. It is desirable that the expected pressure at the nozzle throat be sufficiently lower than the vapor pressure of the solution for the generation of vapor bubbles.
2. In order to maintain the vapor bubble generated at the nozzle throat, it is desirable that the pressure downstream of the nozzle outlet is at least in the metastable region based on the saturated vapor pressure. If the vapor bubble can be maintained downstream of the nozzle, the vapor bubble in the liquid will mostly rise due to buoyancy, and the bubble can grow more easily as it moves to a lower-pressure region.
3. Although the movement of bubbles has not been mentioned yet, it can be seen that the depth of bubbler nozzles for generating vapor bubbles is limited by the preceding two conditions.
4. Although not mentioned yet, in the case of degassing, mass transfer must occur between the vapor bubble and the solution so the depth of the nozzle seems to directly affect not only the generation and maintenance of vapor bubbles but also the deaeration performance.
5. Although not mentioned yet, in the case of desalination applications by generating large-capacity bubbles, additional consideration of the depth of the nozzle seems not critical because the purpose is the phase change of evaporation–condensation rather than mass transfer by vapor bubbles.

2.3 DIFFUSION THROUGH VAPOR BUBBLES

When the pressure inside a vessel containing a liquid is reduced, the solubility decreases according to Henry's law, and as a result, the supersaturated solute is to be released through bubbles. This is a concept that can be applied to the extent that the solution maintains the liquid phase. Simply put, it is effective only if the reduced pressure level is above the saturated vapor pressure of the liquid, which is described as the lower limit of the liquid phase.

In fact, under a given pressure condition ($p > p_{sat}$), bubbles are no more generated once the supersaturated solute in the solution fully escapes from the liquid after a certain period of bubbling. However, when the local pressure is lowered below the saturated vapor pressure of the solvent, vapor bubbles with completely different properties are generated. The conditions for forming bubbles were introduced in the previous section, and now let us talk about the properties of vapor bubbles and their behavior related to the mass diffusion.

2.3.1 COMPOSITION OF VAPOR BUBBLE

What will be the composition inside the vapor bubble? This is an important issue, but it has not been easy to find related literature, because the generation and use of room-temperature vapor bubbles has not been routine until now. On the other hand, high-temperature steam has already been put into practical use for a long time and is already taken for granted, so it may be considered to be of little value to be reported to academia. To conclude, both room-temperature vapor bubbles and high-temperature steam have very low oxygen concentrations. Although we have not confirmed the clear explanatory data that high-temperature steam has a very low oxygen concentration, it is believed that the oxygen concentration of high-temperature steam is already very low in that the final deaeration is completed through steam sparging in the process of thermal deaeration. This may not be entirely new to engineers in this field. In the case of the thermal deaeration process, degassing driven by solubility should occur primarily during the heating process, and the high level of degassing in the final stage is understood to be caused by vapor bubble diffusion through steam sparging. Then, it seems necessary to understand what changes the evaporation process causes in relation to the concentration of oxygen. This can be seen as a common principle applied to both vapor bubbles generated in a vacuum and hot steam generated at high temperatures. For convenience, I will focus my explanation on vapor bubbles in a vacuum state.

First, let us look at a case where water, which was in equilibrium with the atmosphere, is depressurized in a short period of time and reaches a level below the saturated vapor pressure, forming vapor bubbles. This can be understood as a case where cavitation bubbles occur on the propeller suction surface. This is the case introduced in Chapter 1, "Take a Break! 2," from which it could be confirmed that the volume fraction of oxygen inside the vapor is 4.68×10^{-6} or $4.68 \times 10^{-4}\%$, which is very, very low. Now, as a slightly different approach, in "Sidebar 4" of Chapter 2, the pressure of external air containing saturated vapor (for example, a constant partial pressure of water vapor of 3.17 kPa at a temperature of 25°C) is lowered without a change in temperature maintaining equilibrium. Looking at the change in composition inside the bubble from the relationship between the change (decrease) in saturated solubility as the pressure decreases and the vapor pressure of water, as the pressure decreases, the ratio of the saturated vapor pressure to the total pressure increases and the mole fraction of vapor increases, while the mole fractions of oxygen and nitrogen tend to decrease. According to this data (Figure 2.E and Table 2.B), although there are questions about the feasibility and effectiveness of actual implementation, if a saturated vapor pressure (only vapor) of 3.17 kPa exists in the atmosphere at 25°C, complete degassing is possible even under this pressure. Moreover, if the temperature of the surroundings under an equilibrium with the water becomes higher and pure water vapor is available, this can be seen as suggesting that complete degassing may be possible even at higher pressure. That is an interesting story about solubility. Let us leave this discussion on the interpretation of solubility aside and now go into the vapor bubbles.

If vapor bubbles are generated without phase equilibrium, what phenomenon can we expect? Once vapor bubbles are created instantly by decompressing the inside of liquid water that is initially in phase equilibrium with atmospheric pressure, the concentrations of oxygen and nitrogen in the liquid are

$$C_{O_2} = 8.3\,mg\,/\,L = \frac{8.3\times10^{-3}\,g\times1\,mol\,/\,32g}{1L\times10^{-3}\,m^3\,/\,L} = 0.259\,mol\,/\,m^3 \text{ and } C_{N_2} = 13.9\,mg\,/\,L$$

$$= \frac{13.9\times10^{-3}\,g\times1\,mol\,/\,28g}{1L\times10^{-3}\,m^3\,/\,L} = 0.496\,mol\,/\,m^3, \text{ respectively. If this solution is now}$$

reduced to vapor, for example, down to 1 kPa, which is lower than the saturated vapor pressure, the saturated solubility becomes undefinable in Henry's law because the solvent water would no longer exist in a liquid phase. In this case, the volume of each component originally contained in 1 L of water is obtained to give the concentration of each constituent inside the bubble instead of the saturation solubility at the interface:

$$V_{O_2} = \frac{m_{O_2}R_{O_2}T}{p} = \frac{8.3\times10^{-6}\,kg\times0.2598\,kJ\,/\,kg\cdot K\times298.15\,K}{1kPa} = 0.00064\,m^3$$

$$V_{N_2} = \frac{m_{N_2}R_{N_2}T}{p} = \frac{13.9\times10^{-6}\,kg\times0.2968\,kJ\,/\,kg\cdot K\times298.15K}{1kPa} = 0.00123\,m^3$$

$$V_{vap} = \frac{m_{vap}R_{vap}T}{p} = \frac{0.997\,kg\times0.4615\,kJ\,/\,kg\cdot K\times298.15\,K}{1kPa} = 137.183\,m^3$$

Or using the already-available value of specific volume of the vapor under 1 kPa:

$$V_{vap} = m_{vap}v_g = 0.997\,kg\times129.19\,m^3\,/\,kg = 128.8\,m^3$$

From this, the volume fraction of oxygen inside the bubble becomes

$$y_{O_2} = \frac{V_{O_2}}{V_{O_2}+V_{N_2}+V_{vap}} = \frac{0.00064}{0.00064+0.00123+137.183} = 4.6\times10^{-6}.$$

In Chapter 1, "Take a Break! 2," we discussed that the volume fraction of oxygen inside the evaporated water does not differ much, whether it is hot steam at atmospheric pressure or higher or vapor bubbles under a vacuum. But what about the level of oxygen partial pressure in each case? The partial pressure of oxygen in a vapor bubble created in a vacuum will be lower by the ratio of decompression compared to that in hot steam under atmospheric pressure. This may be an important issue in the degassing process, which is governed by concentration difference between phases, especially when the lowest dissolved oxygen concentration level is of concern.

2.3.2 BUBBLE DIFFUSION MODEL

Analysis and prediction tools based on the physical phenomena related to vacuum bubbling that we have examined so far will be a research field that is expected to be in great demand for various industrial applications expected. Here, as a big picture for modeling, we first analyzed and reproduced the discrete bubble model [19] for the aeration process, for which some research results have been reported, and then applied this methodology to the vacuum bubbling deaeration modeling. McGinnis et al. (2002) [19] presented an aeration model using an air bubbler for application in the field of water resources management based on the research results of Wüest et al. [18] and presented a specific methodology in their work. Since this paper describes the entire modeling process in relative detail, there appears to be no particular challenge in performing simple modeling based on it. Here, the results of McGinnis et al.'s research in the field of aeration using the methodology presented are reproduced, and the same theoretical background is applied to vacuum bubbling deaeration. As is clear from the definition of the terminology, in the case of aeration, the direction of mass diffusion is from air bubbles to the liquid, whereas in the case of vapor bubble degassing, the dissolved gas in liquids moves from the liquid to the vapor bubbles. In addition, in the case of aeration, the amount of air supplied is clearly presented, but in the case of degassing, the amount of vapor bubbles generated and the composition of the bubble contents should be known through prior experiments. Since these values are determined according to the bubble generation conditions, there appears to be a continued need for research on establishing bubble generation conditions separately from bubble diffusion model. Other differences between aeration and vacuum bubbling deaeration include the significant growth in bubble size between the time the bubbles are generated and when they reach the water surface, and the internal composition of the generated bubbles continuously changes as the deaeration time elapses. Therefore, modeling is based on the amount of captured gas and dissolved oxygen concentration measured through experiments, through which the initial size of the bubble, SMD, can be estimated.

Vacuum bubbling deaeration is understood as the result of discharge of supersaturated solutes and mass diffusion by vapor bubbles. These two mechanisms can appear separately or simultaneously. By the way, in the former case, degassing progresses quite fast in the initial stage, whereas the latter is a diffusion process that progresses relatively slowly, so detailed understanding and analysis of the latter behavior appears to be more important in understanding the overall behavior of degassing. Here, first, the most appropriate bubble size in terms of Sauter mean diameter (SMD) is estimated based on the degassing experimental data in the late stage, which represents the vapor contribution only, along with the discrete bubble model. Then, all the information necessary for simulating degassing performance, that is, the degassing rate, the composition of gas inside the bubble, and the bubble size, can be determined. The preceding data, although in an early stage, suggest the possibility of simulating the entire degassing process. For reference, in the case of the aeration model, the SMD, which is the average diameter of bubbles based on the surface area, was obtained from measuring the size of the bubbles [19], but in the case of the degassing model, a trial-and-error method was used to find the SMD closest to the experimental data. Now, let us take a look at the discrete bubble model, known as the basic framework for aeration modeling.

2.3.2.1   Discrete Bubble Model Description

The discrete bubble model, first introduced by Wüest et al. [18], has been applied by McGinnis and Little [19] to bubbles that rise in plug flow through a 14 m deep tank of well-mixed water. The assumptions and limitations of the model include the following:

1. The initial bubble size (SMD) and the rate of bubble formation remain constant.
2. Bubble coalescence is negligible, and mass transfer of nitrogen and oxygen is considered.
3. The temperatures of water and air are in equilibrium and remain constant.
4. Mass transfer through the free surface of water is negligible.

The mass transfer flux (subscript i for either oxygen or nitrogen) across the surface of a bubble is:

$$J_i = K_L \left(C_{s,i} - C_i \right) \left(mol \cdot m^{-2} \cdot s^{-1} \right) \tag{i}$$

where K_L is the liquid-side mass transfer coefficient, C_s is the equilibrium concentration at the gas/water interface, and C is the bulk aqueous-phase concentration. The gas-side mass transfer resistance is assumed negligible. The equilibrium concentration from Henry's law is:

$$C_{s,i} = H_i p_i \left(mol \cdot m^{-3} \right), \tag{ii}$$

where H_i is Henry's constant and p_i is the partial pressure of a specific gas at a given depth. Using the expression of Eq. (ii) in the Eq. (i) yields:

$$J_i = K_L \left(H_i p_i - C_i \right) \left(mol \cdot m^{-2} \cdot s^{-1} \right). \tag{iii}$$

The rate of mass transfer for a single bubble is obtained by multiplying the surface area of a bubble of radius r as:

$$\frac{dm_i}{dt} = -K_L \left(H_i P_i - C_i \right) 4\pi r^2 \left(mol \cdot s^{-1} \right) \tag{iv}$$

The vertical location of a bubble is determined by the bubble-rise velocity, v_b, and any induced vertical water velocity, v, by:

$$\frac{dz}{dt} = v + v_b \left(m \cdot s^{-1} \right) \tag{v}$$

where z refers to the vertical coordinate. The induced velocity is assumed low relative to the bubble-rise velocity. A unique feature of this discrete bubble model lies in that instead of conducting time integration on the mass diffusion and the vertical location separately, it combines the two Eqs. (iv) and (v) to give the mass of gaseous species transferred per bubble per unit height of tank as:

$$\frac{dm_i}{dt} = -K_L\left(H_iP_i - C_i\right)\frac{4\pi r^2}{v_b}\left(mol\cdot m^{-1}\right) \tag{vi}$$

through which the variable time t is no longer involved. The number flux of bubbles entering the tank, N, is estimated based on the initial bubble volume, V_0, and the actual volumetric gas flow rate at the diffuser, Q_0, or:

$$N = \frac{Q_0}{V_o}\left(s^{-1}\right) \tag{vii}$$

Multiplying Eq. (vi) by N and expressing it in terms of M, the molar flow rate of gas, yields:

$$\frac{dM_i}{dt} = -K_L\left(H_iP_i - C_i\right)\frac{4\pi r^2 N}{v_b}\left(mol\cdot m^{-1}\cdot s^{-1}\right) \tag{viii}$$

During the time a bubble takes to rise up, the bulk aqueous-phase concentration is assumed to remain constant. Eq. (viii) is integrated numerically, for each species i, to obtain the change in the molar flow rate while the gas bubble is in contact with the water during its rise. Here, H varies with water temperature, while v_b and K_L are functions of r, the radius of the bubble as are given in Table 2.4. According to these data, the bubble-rise velocity changes based on two characteristic sizes (radius), that is, 0.7 mm and 5.1 mm, and the mass transfer coefficient increases linearly below 0.667 mm but is represented by a constant value for radii larger than that. The details of the model, including the temperature dependence of K_L, are introduced in reference [19].

TABLE 2.4
Correlations Used in Discrete Bubble Model (Henry's Constants, Bubble-Rise Velocity, and the Liquid-Side Mass Transfer Coefficient)

Property or Variable Name	Correlation	Range
H_{O_2} $(mol\cdot m^3\cdot bar^{-1})$	$H_{O_2} = 2.125 - 5.021\times10^{-2}T^2 + 5.77\times10^{-4}T^2$ 10^4T^2(T in °C)	Not available
H_{N_2} $(mol\cdot m^3\cdot bar^{-1})$	$H_{N_2} - 1.042 - 2.450\times10^{-2}T^2 + 3.171\times10^{-4}T^2$ 10^4T^2 (T in °C)	Not available
K_L $(m\cdot s^{-1})$	$K_L = 0.6r$	$r < 6.67\times10^{-4}$ m
	$K_L = 4\times10^{-4}$	$r \geq 6.67\times10^{-4}$ m
v_b $(m\cdot s^{-1})$	$v_b = 4474r^{1.357}$	$r < 7\times10^{-4}$ m
	$v_b = 0.23$	$7\times10^{-4} < r < 5.1\times10^{-3}$ m
	$v_b = 4.202r^{0.547}$	$r \geq 5.1\times10^{-3}$ m

Source: [18, 19].

Using these relationships with increasing bubble radius, both the bubble-rise velocity and the mass transfer coefficient are recalculated as the bubble travels up to the free surface. Once the spatial integration is finished when the bubble reaches the surface, the overall changes in the molar flow rate of gas (both oxygen and nitrogen) are used to update the evolving bulk aqueous-phase concentration as the time advances. McGinnis et al. [19] conducted a direct measurement of bubble diameter to obtain SMD in the process of their model validation and used the initial DO concentration, water temperature, and the depth along with the SMD to initialize the model calculation and assumed that the initial dissolved concentrations of both oxygen and nitrogen are at equilibrium with the atmosphere. The initial molar flow rate of gaseous oxygen and nitrogen is:

$$m_0 = \frac{Y_o P_{std} Q_{std}}{RT_{std}} \left(mol \cdot s^{-1} \right) \tag{ix}$$

where Y_o is the initial mole fraction of the gas, P_{std} is the standard pressure, Q_{std} is the gas flow rate at standard temperature and pressure (0°C and 1 bar), R is the universal gas constant, and T_{std} is the standard temperature.

2.3.2.2 Discrete Bubble Model Test Case for Aeration

A test case study to confirm the validity of the preceding discrete bubble model was introduced by McGinnis and Little (2002) [19], and the test conditions they used are summarized in Table 2.5.

In order to verify the validity of the discrete bubble model, McGinnis and Little (2002) [19] conducted measurements of dissolved oxygen concentration at three different airflow rates, that is, 0.43, 0.68, and 2.88 Nm³/h (the normal condition here refers to 0°C and 1 atm), in which a bubbler was installed at a height of 0.6 m from the bottom of a water tank with a diameter of 2 m and a depth of 14 m. The initial bubble size for each of the airflow rates is represented by the measured initial SMD at two different depths of 6.7 and 12.5 m. They mentioned that there was essentially no difference between the bubble sizes formed at these two different depths. Typical SMDs in their experiments are reported to be 1.2 to 1.6 mm, and these bubble sizes and the measured airflow rates were used as input to the discrete bubble model to predict the oxygen concentration as a function of time. They claimed that their predictions fall within 15% of the observed data.

TABLE 2.5
Test Case Description of McGinnis and Little (2002)

Description	Unit	Value(s)	Remarks
Airflow rate	$Nm^3 \cdot h^{-1}$	0.43, 0.68, 2.88	at 0°C, 1 atm
Tank size ($\Phi \times H$)	m	$\Phi 2 \times 14$	
Air diffuser depth	m	13.0	
Water temperature	°C	23	
Probe depth	m	3, 8, 12	

Source: [19].

Figure 2.23 is the experimental result of McGinnis and Little (2002) [19], which shows the change in dissolved oxygen concentration in the water for three supply bubble flow rates (0.4, 0.68, and 2.88 Nm^3h^{-1}) in a tank with a diameter of 2 m and a height of 14 m. The DO concentration results measured at depths of 3, 8, and 12 m did not show a significant difference depending on the measurement depth, and the average value of the experimental data was used for comparison to verify the model (Figure 2.24).

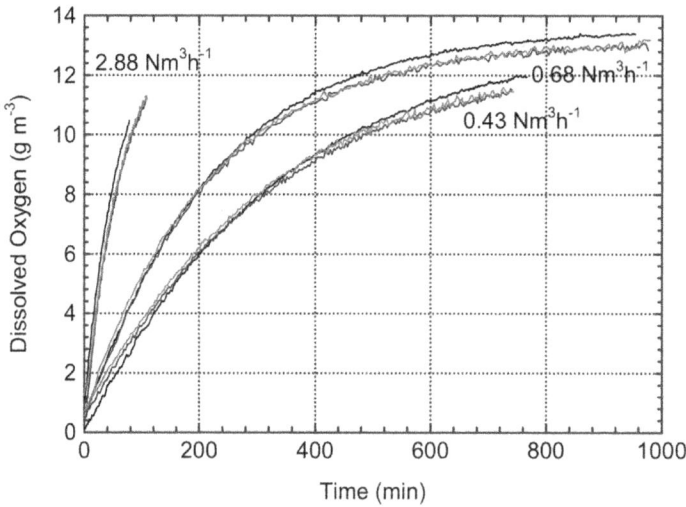

FIGURE 2.23   Mass transfer test results showing measured DO concentrations as a function of time in the process of aeration from McGinnis and Little [19].

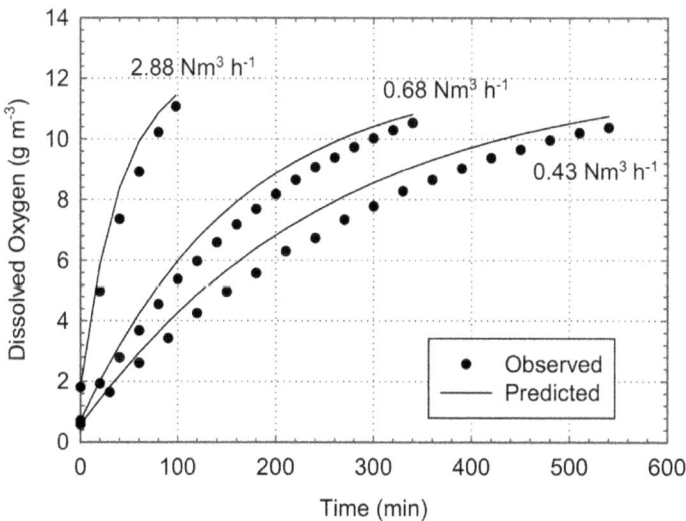

FIGURE 2.24   Averaged observed DO concentration versus the predicted results using the discrete-bubble-model of McGinnis and Little in the process of aeration [19].

2.3.2.3 Numerical Process for Calculations

Eq. (viii), along with the values introduced in Table 2.4, gives the molar diffusion flow rate of gas per depth interval, which are eventually summed up to give the total molar diffusion flow rate of gas in terms of mol/s. Assuming that this flow rate is maintained during the specified time step, total diffused mass from the bubble space to bulk liquid is obtained by multiplying the total molar diffusion flow rate by the specified time step and added to the mol of the dissolved gas in the bulk liquid at previous time step. New concentration is obtained based on the molar mass of a specific gas divided by the total volume of the bulk liquid to give the concentration in terms of mol/m^3 or mg/L. The calculation procedure is illustrated in Figure 2.25.

The input data for this simulation is summarized in Figure 2.26, and the simulation results are shown in Figure 2.27.

McGinnis and Little (2002) [19] applied the SMD value measured as the initial diameter of the bubble in the simulation for comparison with their experimental results. In the simulation shown in Figure 2.27, however, the average diameter that gave results more closely matching the experimental results was found through trial

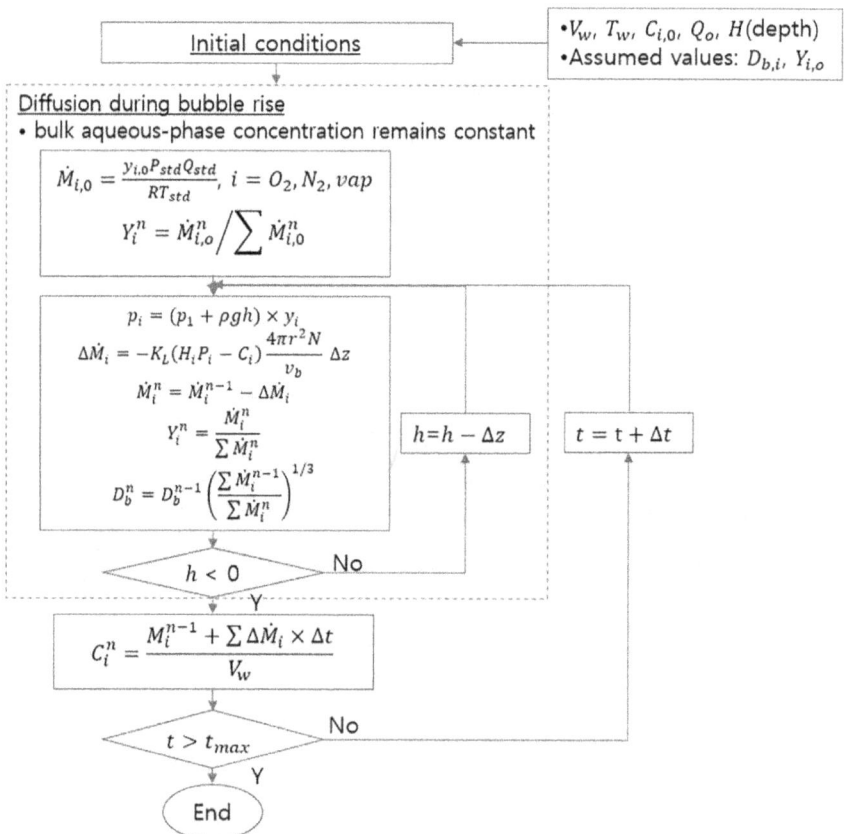

FIGURE 2.25 A flow diagram for aeration process.

- $V_w = \frac{\pi}{4}D^2L = 44m^3$
- $T_w = 23°C$
- $C_{O_2,0} = \frac{1mg}{L} = 0.03125 \ molm^{-3}$ (bulk water)
- $C_{N_2,0} = 14.3\frac{mg}{L} = 0.51071molm^{-3}$ (bulk water)
- $Q_o = 0.43, 0.68, 2,88 \ Nm^3h^{-1}$
- $H_0 = 14$ m
- $D_{b,i} = 1.2 \ mm$
- $y_{O_2,o} = 0.21$ (gas side)
- $y_{N_2,o} = 0.79$ (gas side)

FIGURE 2.26   Simulation input for a validation case of McGinnis and Little [19].

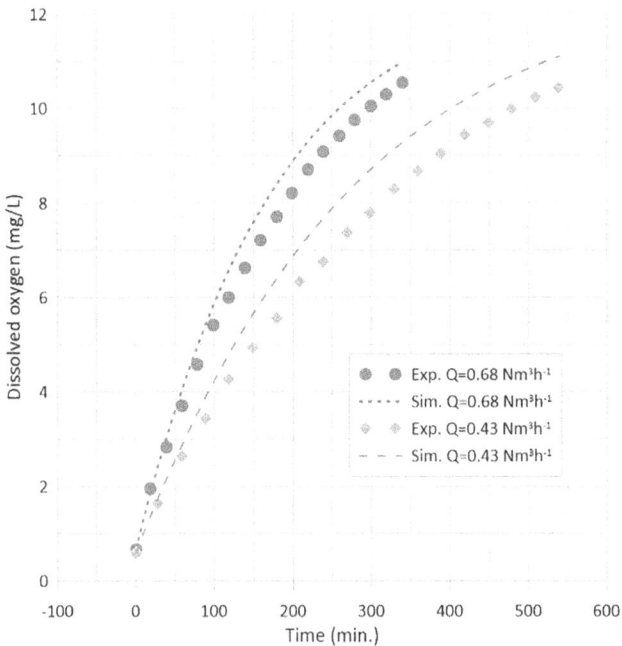

FIGURE 2.27   Aeration test case results compared against the experimental data of McGinnis and Little [19].

and error, and the results shown in the figure are the simulation results when the SMD is 2.0 mm. This is a fairly large difference compared to the SMD = 1.2 mm based on measurements for the initial airflow rates of 0.43 Nm³/h and 0.68 Nm³/h, but it can be seen as a more reasonable reflection of the effects of merging between bubbles during the bubble-rising process. The predicted values of dissolved oxygen concentration shown in Figure 2.27 overestimated the experimental data by up to 15% (flow rate 0.43 Nm³/h) and 10% (flow rate 0.68 Nm³/h), respectively. The calculation based on the *discrete bubble model* described earlier for the case calculation of aeration was actually performed through an Excel worksheet, but the detailed calculation process is introduced through the sample coding image that follows (Figure 2.28). Based on the information presented, it may not be a very challenging task to simulate in the available software.

```
! Sample Coding for Aeration Case Study
! Input Data

Tw = 23.0 ! Water temperature (°C)

pvap = 2.83756 ! Saturated vapor pressure at Tw ← from thermodynamic table

Q0 = 0.68 ! Air flow rate at standard condition of 273.15 K and 101.3 kPa (Nm³/h)

Db,i = 0.0016 ! Initial bubble diameter (m)

p1 = 101.3 ! Ullage pressure (kPa)

D = 2.0 ! Vessel diameter (m)

h = 13 ! Water depth (m)

Co2,im = 0.66667 ! Initial DO concentration in water (mg/L) ← from experiment

Δt = 60 × 5 ! Time step size (sec)

Δh = 0.5 ! Spatial step size (m)


! Related Data and Formula

Ho2 = 2.125 − 5.021 × 10⁻²Tw + 5.77 × 10⁻⁴Tw² ! Henry's constant for oxygen (mole/m³·bar)

HN2 = 1.042 − 2.45 × 10⁻²Tw + 3.171 × 10⁻⁴Tw² ! Henry's constant for nitrogen (mole/m³·bar)

KL = 0.6 × Db/2 ! Liquid-side mass transfer coefficient for Db/2 < 0.667 mm (m/s)

KL = 0.0004 ! Liquid-side mass transfer coefficient for Db/2 ≥ 0.667 mm (m/s)

vb = 4474 × (Db/2)^1.357 ! Bubble rise velocity for Db/2 < 0.7 mm (m/s)

vb = 0.23 ! Bubble rise velocity for 0.7 < Db/2 < 5.1 mm (m/s)

vb = 4.202 × (Db/2)^0.547 ! Bubble rise velocity for Db/2 ≥ 5.1 mm (m/s)

Ru = 8.31447 ! Universal gas constant (J/mol · K)
ρw = 997 ! Water density at Tw (kg/m³)
g = 9.81 ! Gravitational constant (m/s²)
T0 = 273.15 ! Temperature of a standard(N) condition for Q0 (K)
p0 = 101.3 ! Pressure of a standard(N) condition for Q0 (kPa)
! Initial Data and Reduced Data
ti = 0 ! Initial time (sec)

Vb,i = (4π/3)(Db,i/2)³ ! Initial bubble volume (m³)

Vw = (π/4)D²h ! Water volume (m³)

p4 = p1 + ρgh ! Nozzle downstream pressure (kPa)

yo2,i = (p4 − pvap) × 0.209/p4 ! Initial mole fraction of oxygen in bubble air (-)

yN2,i = (p4 − pvap) × 0.791/p4 ! Initial mole fraction of nitrogen in bubble air (-)

yvap,i = pvap/p4 ! Initial mole fraction of vapor in bubble air (-)
```

FIGURE 2.28   A sample illustrative coding for aeration simulation (image).

$\dot{M}_{O_2,i} = \frac{y_{O_2,i} \times p_0 \times Q_0}{R_u T_0}$! Initial number of moles of oxygen in Q_0 (mol/s)

$\dot{M}_{N_2,i} = \frac{y_{N_2,i} \times p_0 \times Q_0}{R_u T_0}$! Initial number of moles of nitrogen in Q_0 (mol/s)

$\dot{M}_{vap,i} = \frac{y_{vap,i} \times p_0 \times Q_0}{R_u T_0}$! Initial number of moles of vapor in Q_0 (mol/s)

$\dot{M}_i = \dot{M}_{O_2,i} + \dot{M}_{N_2,i} + \dot{M}_{vap,i}$! Initial number of moles in Q_0 (mol/s)

$C_{N_2,im} = H_{N_2} \times (1 - p_{vap}/p_1) \times 0.791 \times 28$! Initial DN concentration in water (assumed) (mg/L)

$C_{O_2,i} = C_{O_2,im}/32$! Initial DO concentration in water (mol/m^3)

$C_{N_2,i} = C_{N_2,im}/28$! Initial DN concentration in water (mol/m^3)

$M_{DO_2,i} = C_{O_2,i} \times V_w$! Initial DO moles in water (mol)

$M_{DN_2,i} = C_{N_2,i} \times V_w$! Initial DN moles in water (mol)

! Initialize Iteration

$t = t_i$! Initialize time (sec)

$\dot{M}_{O_2} = \dot{M}_{O_2,i}$

$\dot{M}_{N_2} = \dot{M}_{N_2,i}$

$\dot{M}_{vap} = \dot{M}_{vap,i}$

$Q = \frac{(\dot{M}_{O_2} + \dot{M}_{N_2} + \dot{M}_{vap}) R_u T_w}{p_4}$! Bubble volume flow rate at working condition (m^3/h)

$C_{O_2} = C_{O_2,i}$

$C_{N_2} = C_{N_2,i}$

$M_{DO_2} = M_{DO_2,i}$

$M_{DN_2} = M_{DN_2,i}$

$D_b = D_{b,i}$

DO WHILE $t = t_{max}$

DO WHILE $h \geq 0$

$h = h - \Delta h$! Local depth where bubble is located (m)

$p = (p_1 + \rho g h)/101.3$! Local pressure where bubble is located (atm)

$y_{O_2} = \dfrac{\dot{M}_{O_2}}{\dot{M}_{O_2} + \dot{M}_{N_2} + \dot{M}_{vap}}$

$p_{O_2} = p \times y_{O_2}$

$y_{N_2} = \dfrac{\dot{M}_{N_2}}{\dot{M}_{O_2} + \dot{M}_{N_2} + \dot{M}_{vap}}$

$p_{N_2} = p \times y_{N_2}$

$y_{vap} = \dfrac{\dot{M}_{vap}}{\dot{M}_{O_2} + \dot{M}_{N_2} + \dot{M}_{vap}}$

$p_{vap} = p \times y_{vap}$

FIGURE 2.28 *(Continued)*

$d\dot{M}_{O_2} = -K_L(H_{O_2}p_{O_2} - C_{O_2}) \times 4\pi \left(\frac{D_b}{2}\right)^2 N\Delta h/v_b$! Diffused oxygen per spatial step (mol/s)

$dM_{N_2} = -K_L(H_{N_2}p_{N_2} - C_{N_2}) \times 4\pi \left(\frac{D_b}{2}\right)^2 N\Delta h/v_b$! Diffused nitrogen per spatial step (mol/s)

$\dot{M}_{O_2} = \dot{M}_{O_2} + d\dot{M}_{O_2}$! Number of moles of oxygen in bubbles in nth spatial step (mol/s)

$\dot{M}_{N_2} = \dot{M}_{N_2} + d\dot{M}_{N_2}$! Number of moles of nitrogen in bubbles in nth spatial step (mol/s)

! $\dot{M}_{vap} = \dot{M}_{vap}$ Number of moles of vapor is assumed constant. (mol/s)

$Q = \frac{(\dot{M}_{O_2} + \dot{M}_{N_2} + \dot{M}_{vap})R_u T_w}{p}$! Bubble volume flow rate in nth spatial step (mol/s)

$D_b = 2\left(\frac{3Q}{4\pi N \times 3600}\right)^{1/3}$! Bubble diameter after diffusion from $Q/3600 = \frac{4\pi}{3}\left(\frac{D_b}{2}\right)^3 N$ (m)

END DO ! finish calculating diffusion per spatial step

SUM_$\dot{M}_{O_2}$ = $\sum d\dot{M}_{O_2}$! Total diffused oxygen mass per time step (mol/s)

SUM_$\dot{M}_{N_2}$ = $\sum d\dot{M}_{N_2}$! Total diffused nitrogen mass per time step (mol/s)

$M_{DO_2} = M_{DO_2} + $ SUM_$\dot{M}_{O_2} \times \Delta t$! Total dissolve oxygen in water (mol)

$M_{DN_2} = M_{DN_2} + $ SUM_$\dot{M}_{N_2} \times \Delta t$! Total dissolve nitrogen in water (mol)

$C_{DO_2} = M_{DO_2}/V_w$! Concentration of DO in water (mol/m³) → write C_{DO_2}

$C_{DN_2} = M_{DN_2}/V_w$! Concentration of DN in water (mol/m³) → write C_{DN_2}

$C_{DO_2,m} = C_{DO_2} \times 32$! Concentration of DO in water (mg/L) → write $C_{DO_2,m}$

$C_{DN_2,m} = C_{DN_2} \times 28$! Concentration of DN in water (mg/L) → write $C_{DN_2,m}$

$t = t + \Delta t$! time advancing by Δt.

END DO

FIGURE 2.28   *(Continued)*

2.3.2.4   Application of the Discrete Bubble Model for Deaeration

The discrete bubble model for aeration study may be extended for use in the deaeration process with the direction of mass transfer changed opposite, that is, the dissolved gases in the bulk liquid move into vapor bubbles in the deaeration process, while the non-condensable gases in the air bubble dissolve into the bulk liquid in the aeration process. Initially, water is ideally in a saturated condition under a given temperature and pressure and may contain oxygen and nitrogen that correspond to the saturated condition. Once water is locally evaporated at lowered pressure condition, the concentration of those gases are re-evaluated based on the volume fractions of individual gaseous state, including vapor.

Figure 2.29 illustrates the conceptual models for bubble aeration vs. bubble deaeration. In the case of aeration (a), the saturated oxygen concentration is estimated based on the Henry's solubility law, with the volume fraction of oxygen in the standard air of 20.9%, which is supposedly higher than that of the bulk water, and the mass diffusion from the bubble to the bulk liquid is realized. However, in the case of degassing (b) using vapor bubbles, on the contrary, oxygen dissolved in the bulk liquid is moved to the vapor bubbles by mass diffusion.

The calculation process for deaeration is similar to the aeration case shown in Figure 2.25, but there are some significant differences in the deaeration case. First, the

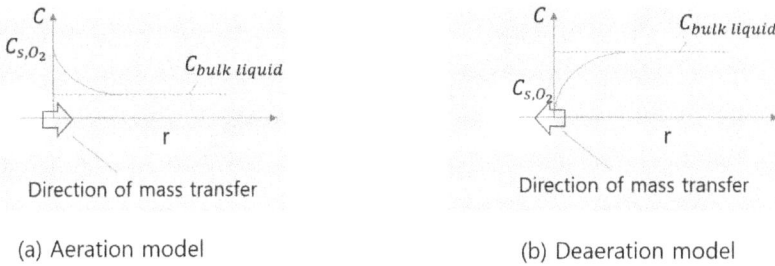

(a) Aeration model

(b) Deaeration model

FIGURE 2.29 Conceptual models for bubble-aeration and bubble-deaeration.

initial conditions for dissolved gas concentration for vapor and bulk liquid are differ-ent. In the case of aeration, there is mass diffusion from air bubbles with high oxygen concentration into the oxygen-depleted liquid (water), as shown in Figure 2.29(a). On the other hand, as shown in Figure 2.29(b), in the case of degassing by vapor bubbles, oxygen moves from a solution (water) in which oxygen is dissolved to vapor bubbles where the oxygen concentration is almost zero. In other words, the direction of mass transfer is opposite. In addition, in the case of aeration as exemplified earlier, a certain amount of air can be supplied through an air pump, whereas in the case of vacuum deaeration, the amount of bubbles generated depends on the internal condi-tions of the liquid. As a means to determine the amount of uncertain vapor bubbles generated, the amount of vapor bubbles during the degassing process could be exper-imentally measured. To validate the bubble diffusion model for outgassing, test cases were selected based on the experimental setup in the Kongju National University laboratory (Figure 2.6 and 2.7), and actual outgassing experiments were performed.

This experiment, which is one of several degassing experiments using vacuum bubbling, is a case where the nozzle depth is located 30 cm below the water surface in a state where the inside of the container is 95% filled with water at room temper-ature (the water volume is about 0.4 m^3). During degassing, the pressure inside the vessel (p_1) was maintained at 1 kPa. Through this experiment, the amount of outgas-sing and the concentration change of dissolved oxygen were measured over time, and the results are presented in Figures 2.30 and 2.31.

The bubble generation rate shown in Figure 2.30 is obtained from the volume of gas collected through the gas collector during degassing. It decreases exponentially from about 2.3 lpm at the beginning and shows an almost-constant bubble generation rate of about 0.034 lpm after about 600 min. This can be seen as the supersaturated solute greatly contributes to bubble generation due to the low container pressure (1 kPa) at the beginning of bubble generation. This phenomenon does not last long, and the amount of bubbles rapidly decreases as the supersaturated solute decreases. The most important fact that can be confirmed through this experimental data is that the final bubble generation rate is maintained at a certain value, not zero. This can be seen as the proof that the bubbles generated are no longer due to solubility but are generated by vaporization of water. It should be noted that the bubbler system used in this experiment is only a selected example, and the degassing rate may vary depend-ing on the system design as well as the thermophysical conditions.

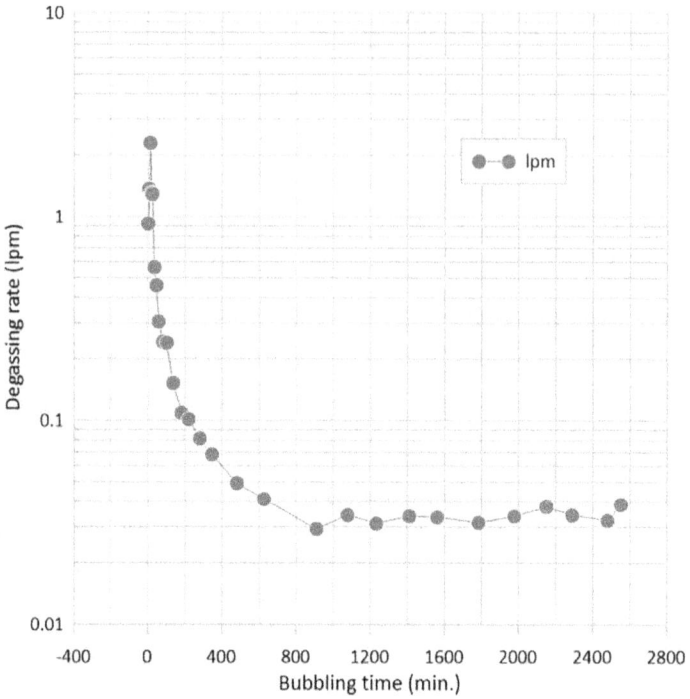

FIGURE 2.30   Degassing rate (lpm) measurement results from vacuum bubbling deaeration ($p_1 = 1\,kPa, \Delta h = 0.3\,m, T_w = 19°C$).

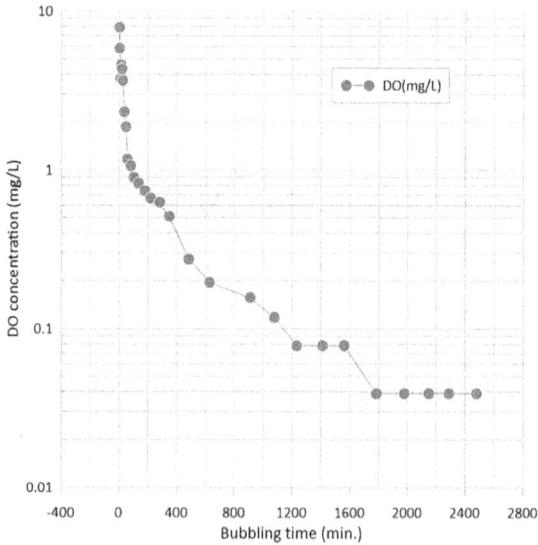

FIGURE 2.31   A measured DO concentration results from a vacuum bubbling deaeration experiment ($p_1 = 1\,kPa, \Delta h = 0.3\,m, T_w = 19°C$).

Figure 2.31 shows the change in dissolved oxygen concentration during degassing. Just as the bubble degassing rate rapidly changed in the early stage of bubbling, the dissolved oxygen concentration decreased very quickly at the beginning and then slowly decreased. This is a result that fits well with the explanation that the supersaturated solute escapes due to the decrease in solubility until a certain point in the early stage of bubble generation, and then the mass diffusion process by vapor bubbles follows. One of the most significant observed results is that the concentration of dissolved oxygen in water continues to decrease, albeit slowly, to a low level of 0.04 mg/L (minimum non-zero reading) at sufficiently low pressure, as can be seen in the figure. (For reference, the values after the last non-zero value are zero but are not displayed in a graph using a logarithmic scale.) This is a result that exceeded the expected lower limit of 0.251mg/L at 20°C, due to the phase change presented earlier, and is therefore of special significance. This also supports the claim that the oxygen concentration inside the vapor bubble is close to zero and is the result of mass diffusion through vapor bubbles. And if you apply this logic, there is theoretically no problem in performing stripping of 5 ppb, the dissolved oxygen concentration requirement known to be achievable through hot steam sparging.

Now let us introduce the modeling of the degassing process. The limitation of simulating the degassing process lies in that the amount of bubbles generated and the initial composition of the gas inside the generated bubbles continuously change as degassing progresses. Considering this, here we would like to first apply the model to the region dominated by vapor bubbles before attempting to simulate the entire bubbling process. In other words, let us first look at the degassing behavior when the generated bubbles are basically vapor bubbles. Figure 2.32 summarizes the initial conditions for simulation of the deaeration test case, and the approximate calculation process of the degassing model by vapor bubbles is illustrated by a flow chart in Figure 2.33.

The test case for calculation uses the degassing rate of the vacuum bubbling degassing process (Figure 2.30) and the dissolved oxygen concentration measurement data (Figure 2.31). From Figure 2.30, the degassing rate gradually decreases and no longer changes, which tells that most of the supersaturated solute has been discharged, and the main composition of the bubbles becomes vapor bubbles. According to Figure 2.30, the corresponding time period appears to be approximately 600 min after the start of bubbling. In this time period, the gas degassing rate no longer changes and shows an almost-constant value (approximately 0.034 lpm) (Figure 2.30), while

- $V_w = 0.65m \times 0.65m \times 0.95m = 0.4m^3$
- $T_w = 20°C$
- $C^0_{O_2} = 0.0109 \ molm^{-3} = 0.35 \ mg/L$ (bulk water)
- $C^0_{N_2} = 0.0207 \ molm^{-3} = 0.58 \ mg/L$ (bulk water)
- $Q_o = 0.00222 \ m^3/h = 0.037 \ lpm$ @20°C
- $H = 0.3 \ m$
- $D_{b,i} = 0.3 \ mm$ (assumed value)
- $Y_{O_2,o} = 3.15 \times 10^{-7} \approx 0.$ (gas side)
- $Y_{N_2,o} = 5.96 \times 10^{-7} \approx 0.$ (gas side)

FIGURE 2.32 Simulation input data for a validation case.

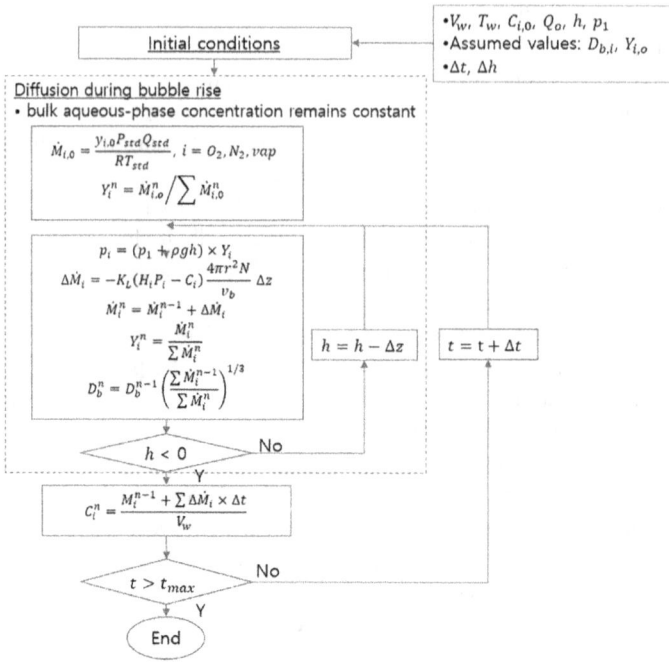

FIGURE 2.33 A flow diagram for deaeration process.

the concentration of dissolved oxygen continues to decrease from 0.2 mg/L to 0.04 mg/L (Figure 2.31)). The first attempt at a degassing model seeks to simulate the time period where the dissolved oxygen concentration is lowered to the lowest limit by vapor bubbles, and the important part here will be not only to confirm the lower limit of dissolved oxygen but also to estimate the rate of decrease over time. This is because, although it is important to check the limit of the degassing level by vacuum bubbling, how quickly degassing can be completed will be an important concern in application.

For modeling purposes, we would like to simulate experimental data after a bubbling time of 482 min. Though the initial gas generation rate slightly deviates from the average value of the subsequent constant flow rate section in this case (Figure 2.30), the rate of change of the dissolved oxygen concentration does not deviate much from that of the subsequent time period (Figure 2.31), so it is intended to utilize more data for comparison. The initial dissolved oxygen concentration of water is approximately 0.27 mg/L, and the bubble generation rate of 0.034 lpm, which is the average value after 600 min, is applied. Since there is no direct measurement value for the concentration of dissolved nitrogen involved in the calculation, it can be arbitrarily estimated at 0.47 mg/L based on the decrease in concentration of dissolved oxygen. For reference, the measured bubble generation rate is converted to a value under the pressure at the location where bubbles emerge during simulation. During the experiment, the bubbler nozzle is installed at a depth of 0.3 m from the water surface, and the vessel pressure is 1 kPa, so the pressure downstream of the nozzle is $p_4 = p_1 + \rho g \Delta h = 1 + 998 \times 9.81 \times 0.3 = 3.94 \, kPa$. For reference, this value is 1.7 kPa higher than the saturated vapor pressure of water at 19°C, 2.2 kPa, and this is not a favorable condition for maintaining the vapor bubbles when they exit the bubbler nozzle. The initial concentration composition inside the bubble for simulation is estimated based on the volume fractions when water with measured dissolved oxygen concentration is vaporized at a specified depth.

Figure 2.30 is the result of dividing the measured volume of gas collected by the elapsed time during the degassing process. However, one of the things to note in vacuum bubbling conditions is that the level of pressure we deal with is so low, about 1/100th of atmospheric pressure, that the effect of hydrostatic pressure is no longer trivial. Now, let us look at the data after the bubbling time of 600 min in the experiment shown earlier to observe the degassing effect by vapor bubbles alone. At this time, the pressure inside the container, p_1, is 1 kPa, but the pressure of the gas contained inside the collector will be determined by the relative depth of the water surface formed inside the gas collector. In the case of this experiment, considering that the depth of the water surface inside the gas collector was, on average, 13 cm lower than the outside of the collector, the pressure of the gas at the time of collection was estimated to be 2.3 kPa. And the average gas generation rate of 0.034 lpm after 600 min shown in Figure 2.30 is converted to that corresponding to the nozzle downstream pressure of $p_4 = 3.93$ kPa and used as the input, 0.021 lpm, of the model calculation. Figure 2.32 summarizes the input data for the degassing model case to verify the degassing effect by vapor bubbles.

One of the issues in simulating the aeration process using the discrete bubble model [18, 19] concerns the bubble size and its distribution. McGinnis and Little (2002) [19] presented results close to the experimental results by using the bubble diameter represented by SMD after measuring the size and distribution of bubbles to discuss this issue. However, the method of actually measuring the size of bubbles and applying SMD by reflecting the distribution results seems not a simple task at all. Rather, finding a diameter close to the measured data under various other conditions may be a more practical application method. In the process of comparing the results of this model with the experimental results (Figure 2.34), it is to be mentioned that the bubble diameter was obtained through this repeated process. In the process of comparing the results of this model with the experimental results (Fig. 2.34), the diameter of the bubble could be estimated through an iterative process. When the degassing flow rate was fixed considering the minute change in pressure between the location of measurement and that of bubble generation, the degassing performance by vapor bubbles was obtained as shown in Figure 2.34, and the corresponding initial SMD of the bubbles was found to be 0.32 mm in the present test case.

Efforts have been made to model the degassing process using vacuum bubbling as a more integrated concept. This is based on the data of several degassing experiments conducted through the vacuum bubbling test apparatus previously introduced. In these experiments, not only the concentration of dissolved oxygen according to the progress of degassing but also the amount of gas collected and the energy (electricity) used during the process were monitored. Among them, data

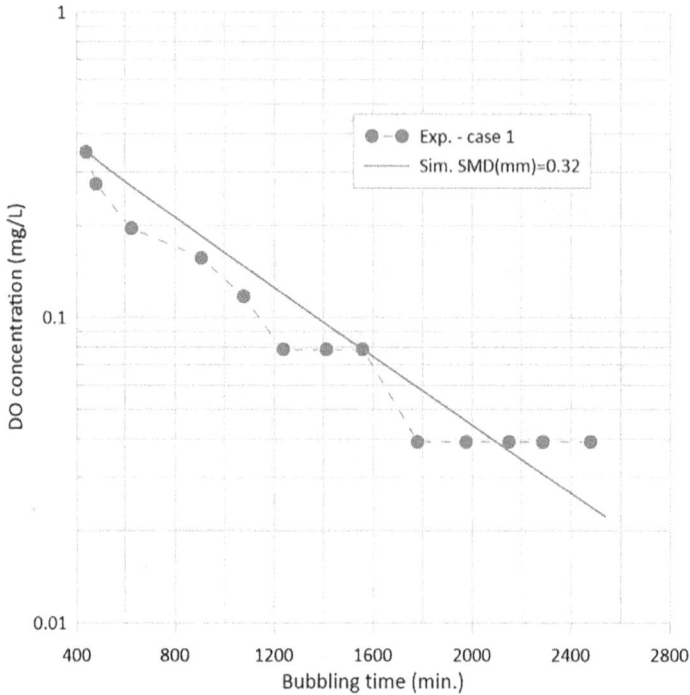

FIGURE 2.34 Deaeration test case results at low DO concentration level below 0.4 mg/L for the test condition specified in Figure 2.32.

on the concentration of dissolved oxygen and the amount of gas collected are to be considered here. The experimental conditions were 95% and 85% loading that correspond to water level of 95 cm and 85 cm in the container, and the pressure was maintained at a minimum of 1 kPa before bubbling but was 2 kPa in most cases due to the pressure rise during gas collection. The water is at room temperature without any external heating, and the temperature varies slightly from experiment to experiment. Though there were temperature increase of about 2–4°C during the bubbling period, the temperature is presented by the initial temperature just for convenience. The depth of the nozzle depends on the loading. In the case of 95% and 85% loading correspond to $h = 0.3$ m and 0.2 m, respectively. The degassing rates are the volume flow rates measured under the experimental condition, and these values are reconverted to the flow rates at the nozzle depth via normal conditions (0°C, 1 atm), reflecting the pressure conditions at the time of measurement. Table 2.6 describes each experimental condition.

Figure 2.35 shows images of vapor bubbles being generated through the bubbler nozzle outlet under the vessel pressure of $p_1 = 1\,kPa$. The main difference between the two conditions is the difference in the depth of the bubbler nozzle from the water surface along with a slight temperature difference. For nozzle depth, case 1 is $\Delta h = 0.3\,m$, and case 2 is $\Delta h = 0.2\,m$. As can be seen in the figures (a) and (b), it can be seen that the bubble plume generated inside the nozzle are maintained

TABLE 2.6
Test Conditions for Vapor Bubbling Deaeration

Test Id.	Nozzle Depth h (m)	p_1 (kPa)	T_w (°C)	Pump Power (W)	p_{sat} (kPa) [31]
Case 1	0.2	1	22	20	2.21
Case 2	0.3	1	19~20	20	2.67

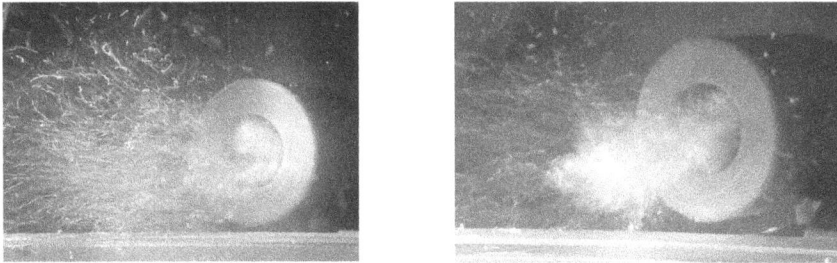

FIGURE 2.35 Typical views of vapor bubble generation ($p_1 = 1\,\text{kPa}$); (a) $\Delta h = 0.3\,\text{m}$, (b) $\Delta h = 0.2\,\text{m}$.

farther downstream of the nozzle. This supports the fact that vapor bubble retention is directly affected by nozzle downstream conditions.

Looking at the change in dissolved oxygen concentration under each test condition in Figure 2.36, it is confirmed that the dissolved oxygen concentration decreased very quickly from the initial dissolved oxygen concentration to reach approximately 1 mg/L at the beginning of bubbling, and after that, it decreased more slowly and finally reached zero (not shown in the figure due the numeric limitation to deal with log scale). The resolution of the dissolved oxygen meter in this experimental device is limited to 0.1% O_2-atm (approximately 0.04 mg/L when converted into a mass unit). The experimental results show that the degassing pattern is largely divided into two stages. The initial phase, in which the rate of reduction is fairly steep (let us call it phase 1 for the sake of discussion), is primarily the phase in which the supersaturated solute escapes. The slow-decreasing phase that follows is the phase in which the remainder of the solute and the vapor bubble work together. At this stage, the amount of remaining solute decreases over time, which eventually is thought to be dominated by vapor bubbles.

This interpretation can be justified to some extent by the data of the degassing rate measured during the bubbling process in Figure 2.37. The outgassing rate can be viewed as the rate of bubble formation, starting with a very large value at the beginning of bubbling and showing a more gradual decrease over time. Looking at the data more carefully, it can be confirmed from the experimental data that the degassing rate does not become zero over time and is maintained at some constant level under these experimental conditions where the pressure inside the container is lower than the vapor pressure of water. These are bubbles that are continuously created regardless

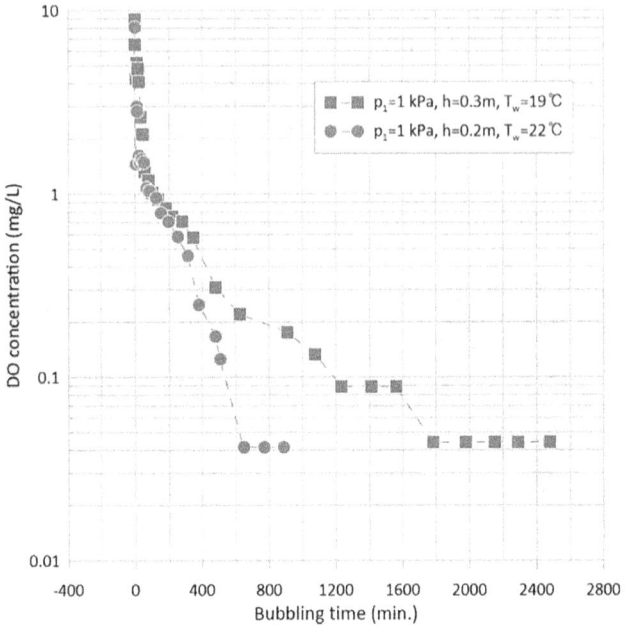

FIGURE 2.36 Measured DO concentrations for different Δh (0.2m and 0.3m) under $p_1 = 1kPa$.

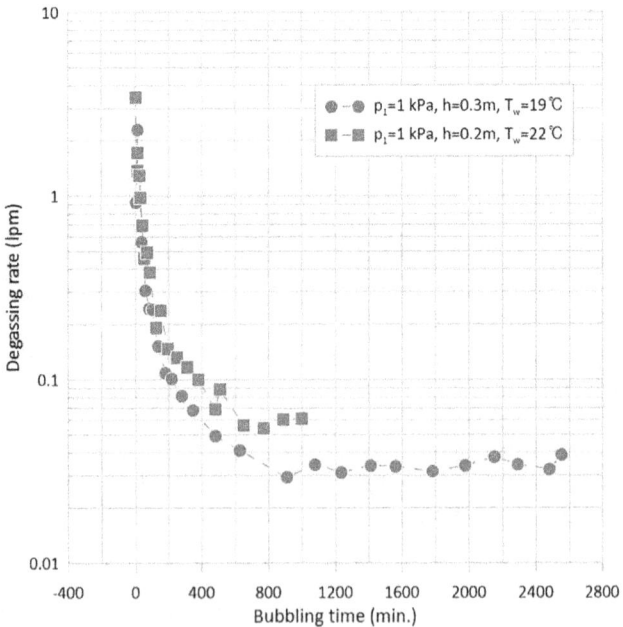

FIGURE 2.37 Measured degassing rates for different Δh (0.2m and 0.3m) under $p_1 = 1kPa$.

of the amount of dissolved gas remaining under that condition, and its substance can be seen as vapor bubble in which water is vaporized. Among the captured volume of extract gas secured by data, vapor bubbles in particular can change to water at any time, depending on the thermodynamic conditions at the location where the bubbles exist, so there is a question in believing that all vapor bubbles generated in the water survive to rise to the surface and are collected. However, it is to be noticed that the vapor bubble generation rate as a measured result varies depending on the operating conditions.

Now, let us take a closer look at the data of Case 1 among the preceding experimental data for modeling for the vacuum bubbling degassing process. As introduced earlier, in the case of aeration modeling, the airflow rate supplied by the pump can be kept constant, but in the case of vacuum bubbling degassing, since bubbles are created inside the liquid (water) according to thermodynamic conditions, information on the amount of bubbles (or degassing rate) and their composition must be provided separately. In addition, when the existing discrete bubble model is applied to the aeration process, only mass transfer by diffusion from supplied air bubbles to water needs to be considered, but in the case of degassing, mass transfer between bubbles and water must be considered in addition to the supersaturated solute gas component already present inside the bubbles when bubbles are created. Now, if we observe the deaeration rate data for Case 1 more carefully, the change in deaeration rate over time can be expressed by a correlation equation as shown in Figure 2.38.

By looking at the degassing behavior, it looks quite clear that the bubble generation behavior is divided into two modes; the first one shows an exponential decrease, while the second one shows almost-constant bubbling rate. This also supports that there exist two modes of bubbling with different characteristics. This is an interesting topic and to be studied further, but at this point, let us make a further assumption on the composition of bubbles at each phase. One possible speculation on the composition may be that the first part of the bubbles is basically a supersaturated solute, as mentioned previously, and the latter a vapor bubble. Actually, this is a very rough assumption, because these two driving mechanism are there all the time and, it would be fair to say, the major mechanism in each phase change from one to another. However, because we have a limited knowledge to say definitely on how bubbles are being initiated under what conditions, I propose a rough model for the composition of the bubbles as follows:

1. Once the bubbling condition is fixed, two modes of bubbling may work simultaneously.
2. Vapor bubble degassing rate is constant, depending on the test condition, and is known or given.
3. Extract gas is composed of vapor bubble as determined in (2), and the rest of the bubbles originate from supersaturated solutes.

Based on the preceding assumption, now the initial composition of the bubbles for simulation based on the discrete bubble model may be provided. Based on the mass balance calculation, the assumed bubble concentration at the time of its generation in the preceding experimental case may be obtained as follows:

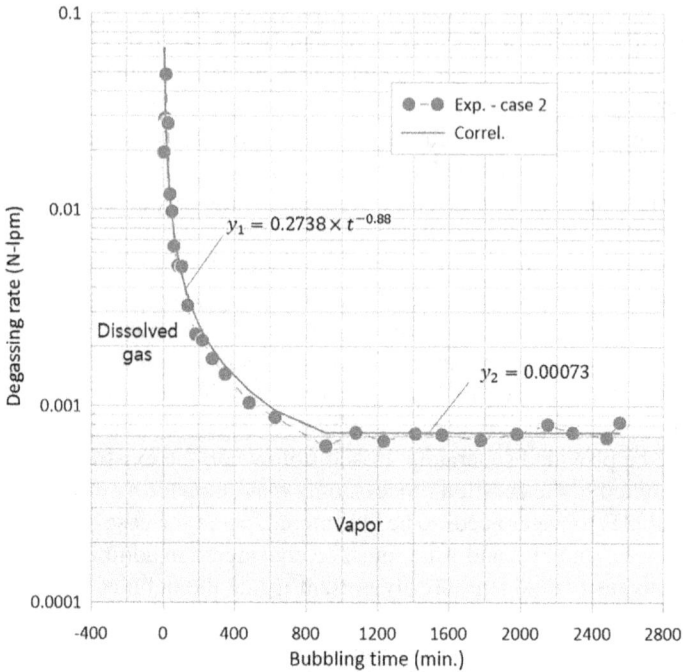

FIGURE 2.38 Two sources of bubbled gas—(1) Solubility-driven dissolved gas, (2) Vapor—under a normal condition.

Calculation of Mass Balance Equation

1. Total volume rate of extract gas data is obtained from experiment (or later predicted). $Q_N(t) in N - lpm$

2. Total volume rate of extract gas data is interpreted to be the sum of vapor bubble and dissolved gas (nitrogen and oxygen). $Q_N(t) = Q_{N,vap}(t) + Q_{N,s}(t)$ Note that $Q_{N,vap}$ (N-lpm) is available from the experimental data, and the volume rate of supersaturated solutes $Q_{N,s}$ is determined.

3. The initial composition of supersaturated solutes is estimated based on the measured DO concentration or based on the assumed equilibrium conditions. $Q_N(t) = Q_{N,vap}(t) + Q_{N,O_2}(t) + Q_{N,N_2}(t)$

   where $\quad Q_{N,O_2} = Q_{N,s} \times 0.332$

   $\qquad\quad Q_{N,N_2} = Q_{N,s} \times 0.636$

4. As time advances, the value of $Q_N(t)$ is provided by the correlated equations with respect to time in minute, as follows (Figure 2.38):

$$Q_N(t) = 0.2738 \times t^{-0.88} \text{ for } 0 < t < t_{phase} = 600\,min$$

$$Q_N(t) = 0.00073 \text{ for } t > t_{phase}$$

Also note that the Henry's solubility law constants for oxygen and nitrogen in water used in this model are as follows [19]:

$$H_{O_2} = 2.125 - 5.021 \times 10^{-2} \times T + 5.77 \times 10^{-4} T^2 \ \left(\text{mol/m}^3 \cdot \text{bar}\right)$$

and

$$H_{N_2} = 1.042 - 2.450 \times 10^{-2} \times T + 3.17 \times 10^{-4} T^2 \left(\text{mol/m}^3 \cdot \text{bar}\right).$$

Based on the preceding assumptions, the volume rates of bubbled gas are shown in Figure 2.39, in which the total and each component of constituent gases are given.

When those gas quantities are used as is in the simulation, the estimated concentration behavior shows reasonable agreement against the experimental results, as shown in Figure 2.40. The trend of the overall concentration change is largely characterized by three zones. The first zone is the one where most of the gases constituting the bubbles are dissolved gases due to the decrease in solubility according to the lowered pressure. In the present test case, up to about 1 hr of the initial bubbling

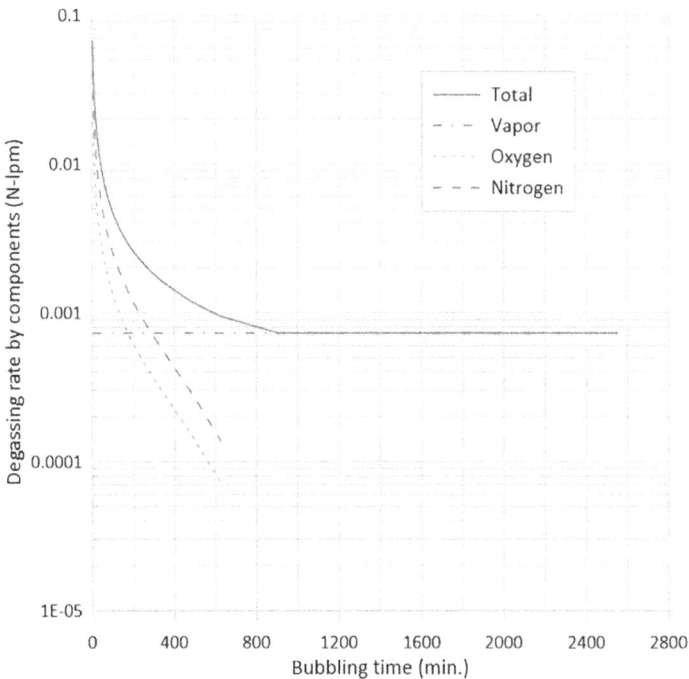

FIGURE 2.39 Contribution of individual gases to global degassing rates under a normal condition.

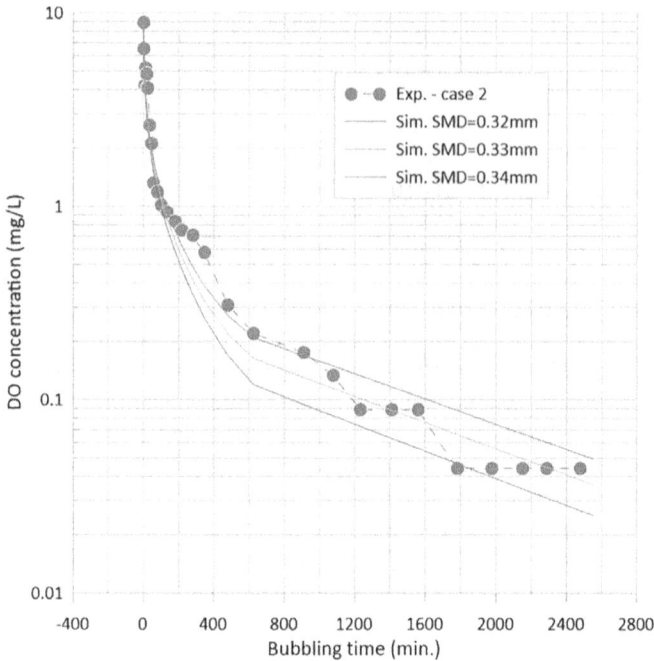

FIGURE 2.40 DO concentration development in water ($\Delta h = 0.3$m, $p_1 = 1 kPa$, $T_w = 19°C$).

corresponds to this. During this time, the concentration decreases very fast, and the dissolved oxygen concentration decreases to about 1 mg/L. In the second zone, the rate of decrease in concentration is somewhat relaxed, and in this case, the relative proportions of dissolved oxygen and vapor bubbles change. This region lasts until the solubility limit (approximately 0.2 to 0.3 mg/L based on the experimental data) set by the phase change. The third zone is dominated by mass transfer between the vapor bubbles and the remaining dissolved oxygen in the water. In this zone, where the contribution of vapor bubbles is dominant, the concentration decrease rate is lower than in the previous zones, but the lowest level of dissolved oxygen concentration can be achieved through continuous operation. In this modeling, (1) the SMD of the vapor bubbles was estimated based on the concentration change due to vapor bubbles of constant rate, and then this bubble size was applied to the entire degassing process, and (2) the collected gas is assumed to be composed of constant rate of vapor and time-dependent contribution from supersaturated solutes. Although these two important yet rigorous assumptions were applied, the simulation results appear to be meaningful in that they reproduce the experimental results relatively well. In Figure 2.40, the initial SMD of bubbles is assumed to be 0.32–0.34 mm. It is to be commented that the final value of the read concentration in the present test case was 0 (zero), which cannot be included in a log scale graph, and the resolution of the measuring device was 0.04 mg/L. The simulation results shown in Figure 2.40 appear to be meaningful in that they reproduce the entire process from the beginning to the end of vacuum bubbling based on experimental data.

The simulated DO concentration describes the overall behavior of concentration development of DO from the initial stage to the later part of degassing through diffusion. So far, the degassing experiment data by vapor bubbles have been reproduced through theoretical modeling based on the previous hypotheses, and the feasibility has been examined. Although this method has limitations in not including detailed information such as the size distribution of generated bubbles according to pressure conditions, it is characterized by being based on the measured degassing rate and dissolved oxygen concentration in water. However, designing new systems requires further modeling efforts. In other words, degassing rate information according to the level of dissolved oxygen over time is required in order to be applicable modeling even when the capacity is changed. To this end, after obtaining an approximate relational expression through regression analysis of the experimental data, the degassing rate information according to the concentration is obtained from the relationship between these approximated formulas. A conceptual introduction to this process is shown in Figure 2.41, which is based on the hypothesis that the outgassing rate depends on the amount of dissolved gas remaining and not as a function of time.

The approach presented in Figure 2.41 is a method of inferring a correlation equation based on the experimental results and using it for simulation. First, data on the change in dissolved oxygen concentration and degassing rate (bubble collection rate) are obtained through a vacuum bubbling experiment as shown in Figure 2.41(a). Next is the process of deriving regression equations for the experimental data. There may be other approaches to this process, but here, we first obtained a regression equation for degassing rate versus time, as shown in Figure 2.41(b), and converted it to normal

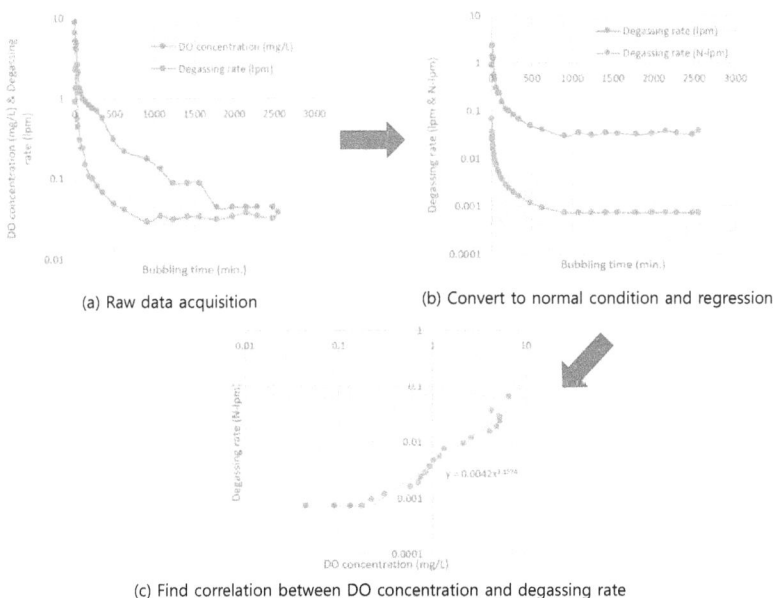

(a) Raw data acquisition

(b) Convert to normal condition and regression

(c) Find correlation between DO concentration and degassing rate

FIGURE 2.41 Modeling process to extract relationship between the DO concentration and degassing rates.

conditions (0°C and 1 atm). Finally, we derive the correlation between degassing rate and dissolved oxygen concentration as shown in Figure 2.41(c). This approach of linking dissolved oxygen concentration directly to degassing rate is quite a risk, but if it can be made feasible, it would represent a big step forward in obtaining predictive tools. Following the aforementioned process, the predicted result of the dissolved oxygen concentration obtained by the experiment is shown in Figure 2.42. As a result of this simulation, we confirmed that similar results to Figure 2.40 could be obtained. The prediction results assuming an average bubble diameter of SMD = 0.32 mm were somewhat conservative compared to the results in Figure 2.40, but the validity of the model could be confirmed to some extent from these results. Although this result is still in an early stage, it can be seen as heralding the emergence of an experiment-based prediction tool in the field of vacuum bubbling deaeration.

So far, a performance prediction method for vacuum bubbling deaeration using bubbler nozzle has been discussed. Generating large amounts of vapor bubbles in a vacuum state and allowing them to pass through the liquid in the form of bubbles may serve as an ideal model in applications requiring high levels of degassing. This is because vapor bubbles have almost zero oxygen concentration, which is very advantageous for mass diffusion, and if the bubble shape is maintained, a significant increase in surface area is also expected while rising up through the liquid. In order to secure optimal operating conditions, various performance design parameters, such as water temperature, vessel pressure, nozzle depth, and power required for nozzles and pumps, should be arranged accordingly.

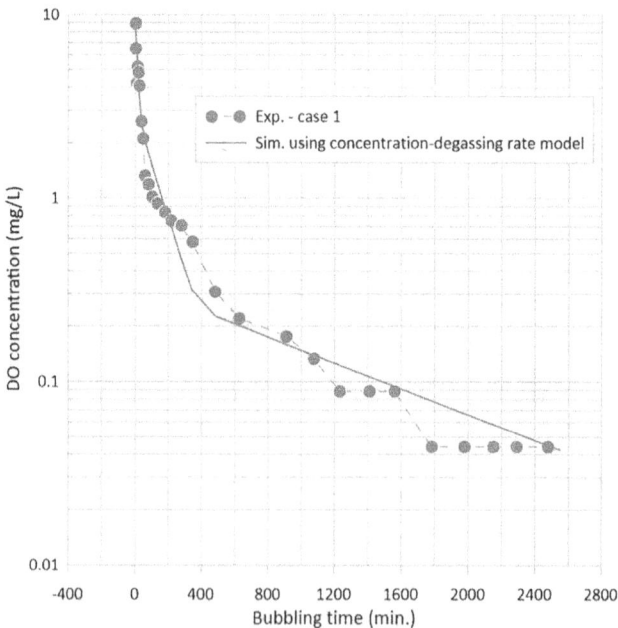

FIGURE 2.42   Estimated DO concentration development using the concentration-degassing rate correlation model.

In this section, four important design parameters for the generation of vapor bubbles during degassing using a venturi bubbler, an example case of vacuum bubbling, were introduced. However, since the mass transfer by the generated vapor bubbles is directly related to the surface area of the bubbles and the residence time in water, it is up to the designer in the relevant field to prepare the optimal design conditions according to the application field. Since the author does not yet have experience in research on all variables, it is not appropriate to mention all expected situations, but some cautions for future designers to refer to are as follows.

1. *Decompression conditions of the container (p_1).* Through experiments so far, when using a rotary vacuum pump with a capacity of 0.75 kW, the vacuum level at room temperature was limited to about 1 kPa. Since this pressure level is lower than the saturated vapor pressure of water at room temperature (3.17 kPa at 25°C), at least the water surface or a space down to a depth of about 20 cm below the water surface belongs to the metastable region from a thermodynamic point of view. However, the lowest level of reduced pressure that can be realized by the vacuum pump may change according to the temperature of the water. As the temperature of the water rises, the evaporation rate increases, and the level of the vacuum pressure that achieves equilibrium may also change accordingly. However, it is not yet clear whether the minimum achievable vessel pressure that changes with temperature is governed by properties such as the surface evaporation rate of the liquid or by the capacity of the vacuum pump or by both.

2. *Depth of bubbler nozzle (Δh).* The depth of the bubbler nozzle acts as a defining factor that determines the destiny of vapor bubbles at the nozzle downstream due to the added hydrostatic pressure contribution to the vessel pressure p_1. The pressure at the nozzle downstream p_4 is $p_4 = p_1 + \rho g \Delta h$, and if this value is greater than the saturated vapor pressure p_{sat} of water, the vapor bubbles created through the nozzle will shrink or disappear. When p_4 is less than p_{sat}, there is a possibility that the bubble exiting the nozzle outlet further grows. It is to be noted that the contribution of the hydrostatic pressure is significant in the case of a vacuum state, unlike under atmospheric pressure. For example, under the aforementioned experimental conditions with $\Delta h = 0.3m$ ($p_1 = 1kPa$, $p_4 = 3.94kPa$), the pressure ratio of the pressure downstream of the nozzle to the water surface is $p_4:p_1 = 3.94:1$, and the volume ratio of bubbles ($V_4:V_1 = 1:3.94$), bubble diameter ratio ($D_4: D_1 = 1:\sqrt[3]{3.94} = 1:1.58$), and bubble surface area ratio ($A_4:A_1=1:2.49$). It is expected that the optimal designs for various application fields considering the conditions for the generation and retention of vapor bubbles will soon appear.

3. *Bubbler performance ($p_4 - p_3$).* Bubbler performance, $p_4 - p_3$, is related to the conditions for generation and retention of vapor bubbles and should be designed properly, which includes the nozzle configuration design in combination with the pump power. It should be remembered that if this

value is kept high, though, not only more power is consumed, but also the problem of airtightness may arise due to expansion–contraction of the gas due to frictional heat when operating in liquid.

The calculation using the discrete bubble model for deaeration case calculation was actually performed through an Excel worksheet, but the detailed calculation process was introduced through the sample coding that follows (Figure 2.43). Based on the information presented, it should not be too difficult to perform a simulation using an available software tool.

! Sample Coding for Deaeration Case Study

! Input Data

$T_w = 19.0$! Water temperature (°C)

$p_{vap} = 2.2125$! Saturated vapor pressure at T_w ← from thermodynamic table

$Q = 0.037$! Vapor flow rate at T_w and 101.3 kPa (lpm) ← from measurement

$D_{b,i} = 0.004$! *Initial bubble diameter (m)*

$p_1 = 1$! Ullage pressure (kPa)

$V_w = 0.401$! Water volume (m³)

$h = 0.3$! Water depth (m)

$\Delta t = 60 \times 1$! Time step size (sec)

$\Delta h = 0.02$! Spatial step size (m)

! Related Data and Formula

$H_{O_2} = 2.125 - 5.021 \times 10^{-2} T_w + 5.77 \times 10^{-4} T_w^2$! Henry's constant for oxygen (mole/m³·bar)

$H_{N_2} = 1.042 - 2.45 \times 10^{-2} T_w + 3.171 \times 10^{-4} T_w^2$! Henry's constant for nitrogen (mole/m³·bar)

$K_L = 0.6 \times \frac{D_b}{2}$! Liquid-side mass transfer coefficient for $\frac{D_b}{2} < 0.667\ mm$ (m/s)

$K_L = 0.0004$! Liquid-side mass transfer coefficient for $\frac{D_b}{2} \geq 0.667\ mm$ (m/s)

$v_b = 4474 \times \left(\frac{D_b}{2}\right)^{1.357}$! Bubble rise velocity for $\frac{D_b}{2} < 0.7\ mm$ (m/s)

$v_b = 0.23$! Bubble rise velocity for $0.7 < \frac{D_b}{2} < 5.1\ mm$ (m/s)

$v_b = 4.202 \times \left(\frac{D_b}{2}\right)^{0.547}$! Bubble rise velocity for $\frac{D_b}{2} \geq 5.1\ mm$ (m/s)

$R_u = 8.31447$! Universal gas constant (J/mol · K)

$\rho_w = 997$! Water density at T_w (kg/m³)

$g = 9.81$! Gravitational constant (m/s²)

! Initial Data and Reduced Data

$t_i = 0$! Initial time (sec)

$V_{b,i} = \frac{4\pi}{3}\left(\frac{D_{b,i}}{2}\right)^3$! Initial bubble volume (m³)

$p = p_1 + \rho g h$! Nozzle downstream pressure (kPa)

$C_{s,O_2} = H_{O_2} \times \frac{(p_4 - p_{vap})}{101.3} \times 0.209/34738$! Initial oxygen concentration in vapor bubble (mol/m³)

$C_{s,N_2} = H_{N_2} \times \frac{(p_4 - p_{vap})}{101.3} \times 0.791/34738$! Initial nitrogen concentration in vapor bubble (mol/m³)

$C_{vap} = \frac{p_4 \times 1}{R_u T}$! Initial vapor concentration in vapor bubble (mol/m³)

FIGURE 2.43 A sample illustrative coding for degassing simulation (image).

$$y_{O_2} = \frac{M_{O_2}}{M_{O_2} + \dot{M}_{N_2} + \dot{M}_{vap}}$$

$$y_{N_2} = \frac{M_{N_2}}{M_{O_2} + \dot{M}_{N_2} + \dot{M}_{vap}}$$

! $C_{DO_2,im}$! Initial DO concentration in water (mg/L) ← from measurement

$C_{DN_2,im} = H_{N_2} \times (p_4 - p_{vap})/101.3) \times 0.791 \times 28$! Initial DN concentration in water (guessed) (mg/L)

$C_{O_2,i} = C_{O_2,im}/32$! Initial DO concentration in water (mol/m³)

$C_{N_2,i} = C_{N_2,im}/28$! Initial DN concentration in water (mol/m³)

$M_{DO_2,i} = C_{O_2,i} \times V_w$! Initial DO moles in water (mol)

$M_{DN_2,i} = C_{N_2,i} \times V_w$! Initial DN moles in water (mol)

$d\dot{M}_{O_2} = K_L(H_{O_2}p_{O_2} - C_{O_2}) \times 4\pi \left(\frac{D_b}{2}\right)^2 N\Delta h/v_b$! Diffused oxygen per spatial step (mol/s)

$d\dot{M}_{N_2} = K_L(H_{N_2}p_{N_2} - C_{N_2}) \times 4\pi \left(\frac{D_b}{2}\right)^2 N\Delta h/v_b$! Diffused nitrogen per spatial step (mol/s)

$M_{O_2} = \dot{M}_{O_2} + d\dot{M}_{O_2}$! Number of moles of oxygen in bubbles in nth spatial step (mol/s)

$M_{N_2} = \dot{M}_{N_2} + d\dot{M}_{N_2}$! Number of moles of nitrogen in bubbles in nth spatial step (mol/s)

$M_{vap} = Q \times C_{vap}$ Number of moles of vapor is assumed constant. (mol/s)

$Q = \frac{(M_{O_2}+M_{N_2}+M_{vap})R_uT_w}{p}$! Bubble volume flow rate in nth spatial step (mol/s)

$D_b = 2 \left(\frac{3Q}{4\pi N \times 3600}\right)^{1/3}$! Bubble diameter after diffusion ← $Q/3600 = \frac{4\pi}{3}\left(\frac{D_b}{2}\right)^3 N$ (m)

! Iteration begins.

DO WHILE t = t_{max}

DO WHILE h ≥ 0

h = h − Δh ! Local depth where bubble is located (m)

p = $(p_1 + \rho gh)/101.3$! Local pressure where bubble is located (atm)

$C_{s,O_2} = \frac{M_{O_2}}{\left(\frac{Q}{3600}\right)}$! Oxygen concentration in vapor bubble (mol/m³)

$C_{s,N_2} = \frac{M_{N_2}}{\left(\frac{Q}{3600}\right)}$! Nitrogen concentration in vapor bubble (mol/m³)

$C_{vap} = \frac{M_{vap}}{\left(\frac{Q}{3600}\right)}$! Vapor concentration in vapor bubble (mol/m³)

$$y_{O_2} = \frac{M_{O_2}}{M_{O_2} + \dot{M}_{N_2} + \dot{M}_{vap}}$$

$$y_{N_2} = \frac{M_{N_2}}{M_{O_2} + \dot{M}_{N_2} + \dot{M}_{vap}}$$

$$y_{vap} = \frac{M_{vap}}{M_{O_2} + \dot{M}_{N_2} + \dot{M}_{vap}}$$

FIGURE 2.43 *(Continued)*

$dM_{O_2} = -K_L(H_{O_2}p_{O_2} - C_{O_2}) \times 4\pi \left(\frac{D_b}{2}\right)^2 N\Delta h/v_b$! Diffused oxygen per spatial step (mol/s)

$d\dot{M}_{N_2} = -K_L(H_{N_2}p_{N_2} - C_{N_2}) \times 4\pi \left(\frac{D_b}{2}\right)^2 N\Delta h/v_b$! Diffused nitrogen per spatial step (mol/s)

$M_{O_2} = \dot{M}_{O_2} + d\dot{M}_{O_2}$! Number of moles of oxygen in bubbles in nth spatial step (mol/s)

$\dot{M}_{N_2} = \dot{M}_{N_2} + d\dot{M}_{N_2}$! Number of moles of nitrogen in bubbles in nth spatial step (mol/s)

! $\dot{M}_{vap} = \dot{M}_{vap}$ Number of moles of vapor is assumed constant. (mol/s)

$Q = \frac{(M_{O_2} + M_{N_2} + M_{vap})R_u T_w}{p}$! Bubble volume flow rate in nth spatial step (mol/s)

$D_b = 2\left(\frac{3Q}{4\pi N \times 3600}\right)^{1/3}$! Bubble diameter after diffusion from $Q/3600 = \frac{4\pi}{3}\left(\frac{D_b}{2}\right)^3 N$ (m)

END DO ! finish calculating diffusion per spatial step

SUM_$\dot{M}_{O_2} = \sum d\dot{M}_{O_2}$! Total diffused oxygen mass per time step (mol/s)

SUM_$\dot{M}_{N_2} = \sum d\dot{M}_{N_2}$! Total diffused nitrogen mass per time step (mol/s)

$M_{DO_2} = M_{DO_2} + $ SUM_$\dot{M}_{O_2} \times \Delta t$! Total dissolve oxygen in water (mol)

$M_{DN_2} = M_{DN_2} + $ SUM_$\dot{M}_{N_2} \times \Delta t$! Total dissolve nitrogen in water (mol)

$C_{DO_2} = M_{DO_2}/V_w$! Concentration of DO in water (mol/m³) → write C_{DO_2}

$C_{DN_2} = M_{DN_2}/V_w$! Concentration of DN in water (mol/m³) → write C_{DN_2}

$C_{DO_2,m} = C_{DO_2} \times 32$! Concentration of DO in water (mg/L) → write $C_{DO_2,m}$

$C_{DN_2,m} = C_{DN_2} \times 28$! Concentration of DN in water (mg/L) → write $C_{DN_2,m}$

$t = t + \Delta t$! time advancing by Δt.

END DO

FIGURE 2.43   *(Continued)*

2.4   MASSIVE EVAPORATION

So far, a method to change the phase of a material from liquid to vapor by reducing pressure, that is, the conditions and methodologies to generate vapor bubbles, has been introduced. To obtain a vapor bubble by decompression from a liquid, the pressure must be reduced below the saturated vapor pressure, where it belongs to the metastable region. The theoretical background, experimental results, and prediction model were presented for a chosen method, an approach to create and maintain vapor bubbles through a venturi nozzle bubbler under vacuum conditions. As an extension of this discussion, then, what is the limit of vapor bubble generation that we can get, and how can we achieve it? If feasible, what are the implications of this?

If the liquid exists locally in a metastable region, what are going to be the additional conditions for the actual phase change to occur, because the necessary conditions for the phase change have already been met in terms of the thermodynamic states? Additional conditions for phase change can be viewed as phase change energy or vaporization energy, and there are various ways to supply vaporization energy, including heating and pressure reduction (mechanical work). In other words, (local) heating and (local) decompression are two typical methods of supplying the energy

required for phase change when the liquid is in a phase-change condition or metastable region and the options for the methods seem to be completely open. In thermodynamics, matter has a unique energy level due to its thermodynamic state, and in
the case of water, the energy levels of liquid and gas states along with the energy for
phase change are listed in most thermodynamics textbooks, such as [23]. The energy
involved when the phase changes from liquid to vapor is called heat of fusion and is
expressed in terms of enthalpy. The data in Figure 2.44 shows the enthalpy of phase
change h_{fg} for water as a function of temperature [31].

Figure 2.44 is the enthalpy data for the saturated liquid (water), and the graphs
show the enthalpy h_f of the liquid (water), the enthalpy of phase change h_{fg}, and the
enthalpy of vapor h_g according to the temperature change. As is clear from the figure,
the enthalpy of the vapor state is the sum of the enthalpy of the liquid phase and the
enthalpy of the phase change. Let us take a look at the data corresponding to 0°C to
100°C in the figure, noting that this is the temperature range at which water exists
as a liquid under atmospheric pressure. The enthalpy required for heating the liquid,
that is, the specific heat, is almost constant, and the enthalpy increases linearly. On
the other hand, the phase-change enthalpy decreases as the temperature increases,
and as a result, the enthalpy of the vapor phase, which is the sum of these two values,
shows a gentle increase with the increase in temperature. This data tells us that when
water is to be vaporized, the amount of energy required to vaporize either at a high

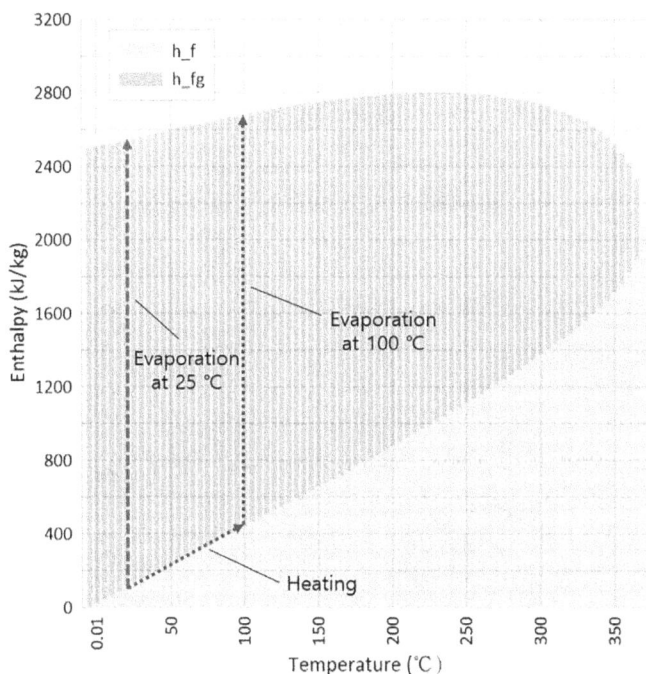

FIGURE 2.44 Enthalpy of saturated water [31].

temperature or at room temperature seems not to make a significant difference. For example, the energy required to vaporize water at 25°C after heating up to 100°C is $h_{g,100°C} - h_{f,25°C} = (2,675.6 - 104.8)$ $kJ/kg = 2,570.8$ kJ/kg, whereas the energy required to vaporize water at 25°C without heating becomes $h_{g,25°C} - h_{f,25°C} = (2,546.5 - 104.83)$ $kJ/kg = 2,441.7$ kJ/kg, resulting in the difference of only about 5.4%. From this, it can be confirmed that the supply of a considerable amount of vaporization energy is essential to vaporize water regardless of heating.

If so, what are the advantages of vaporizing water at a low temperature (or room temperature) compared to the existing method of vaporizing after heating to a high temperature? Although the energy essential for vaporization is absolutely necessary, the energy source does not necessarily have to be a high-temperature heat source, such as the combustion of fossil fuels, if vaporization can be performed at a lower temperature. In other words, if the energy for heating and evaporation can be supplied without consuming fossil fuels, the demand for massive use of fossil fuel can be reduced. In this case, the necessary low-level energy may be supplied and utilized through local renewable energy, for example. An application area for which this scenario is particularly suitable would be desalination, where the current high-energy consumption limits the expansion of the application.

As shown in Figure 2.45, let us assume that the state of the liquid reached the region of saturated vapor pressure or lower (metastable region) on the phase-change diagram through depressurization. Since obtaining vapor rather than bubbles is of greater interest here, heating, decompression, or a combination of these may be considered as possible ways to obtain vapor. How to obtain vapor is no longer a theoretical issue but is simply a matter of choice, depending on the actual application situation, taking into account of economic feasibility, environmental impact, initial investment cost, the complexity of operation, etc.

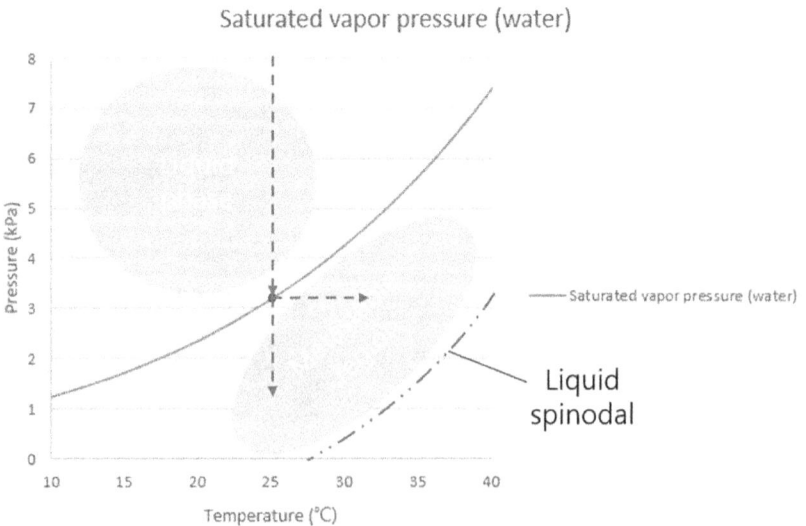

FIGURE 2.45 Possible approaches for massive evaporation at a saturated vapor pressure condition.

For convenience of discussion, if the thermodynamic state of the liquid is under the phase-change condition as shown in Figure 2.45 (e.g., $T_w = 25°C$, $p = p_{sat} = 3.17\,kPa$), the thermodynamic condition for vapor bubble generation is already satisfied at least state variable–wise. So if there is an additional energy supply, this energy could be used to create vapor bubbles. At this time, the amount of vapor bubbles created can be expected to be generated in proportion to the additionally supplied energy or mechanical work. If the energy applied to the system δQ is solely used for vaporizing water, the amount of water that can be vaporized per unit energy input (kWh) is 0.39 g/kJ or 1.40 kg/kWh, noting that $h_{fg@25°C} = 2570.8\,kJ\,/\,kg$. Let us assume that the energy required for vaporization is supplied by electrical energy through a solar PV system and the conversion efficiency from electrical energy to thermal energy is 1. Assuming that a 3 kW home solar power house module is used and the peak time is 5 hr, the total amount of electricity that can be produced per day and the expected evaporation capacity are $E_{elec} = 3\,kW \times 5\,hr\,/\,day = 15\,kWh\,/\,day$ and $m_{evap} = 1.40\,kg\,/\,kWh \times 15\,kWh\,/\,day = 21\,kg\,/\,day$. Of course, since this process assumes a state in which a reduced pressure is maintained, the driving power of the vacuum pump to maintain the reduced pressure must be added, and additional considerations such as energy conversion efficiency will be required. Therefore, this result can be regarded as the best potential performance expected when using solar energy for vaporization. As examples of practically commercialized vacuum evaporation, a local enhanced evaporation method using a packing material and a method of enhancing evaporation by maximizing the surface area have already been applied but are limited to product development mainly for deaeration applications. Evaporation by local direct heating in a saturated state under reduced pressure seems to be the most direct and efficient technology, but no commercialized case has been identified yet. In this section, we will briefly introduce a method for generating vapor or low-temperature steam by operating a pump using electrical energy.

Desalination is a possible application field to which vacuum massive evaporation may be applied. The traditional method of desalination is to boil water to make water vapor, then cool the steam again to obtain condensate. Or it is that water is separated from other minerals or dissolved gases by a reverse osmosis (RO) method using a concentration difference through a membrane. As is known, there are cases where vacuum is used in membrane systems for desalination, but it is a different approach from the vacuum bubbling method, and the possibility of desalination through the vacuum bubbling method can be summarized as follows:

If there is an available energy source such as solar energy for vacuum bubbling, a certain degree of temperature increase can help alleviate the conditions for phase change. In the case of vacuum bubbling, if the water temperature is raised to a certain extent (for example, to 40°C), more relaxed operating conditions and a greater amount of vaporization are expected due to an increase in saturated vapor pressure. Since the temperature rise of this degree is significantly lower than that of the evaporation method by heating (about 100°C or more), it is expected that this level of temperature increase can be achieved by heating without a combustion process from low-grade energy sources rather than heating with fossil fuels. Realizing phase change through the mechanical work of vacuum bubbling at this temperature is of great significance in that it replaces the core process of desalination with renewable energy such as solar energy without burning the fossil fuel required for the process.

Vacuum bubbling in the deaeration or desalination process differs from vacuum-packed bed, another vacuum application. In the case of the latter, there is an advantage in that the liquid can be evaporated more easily using surface modification in the process of surface evaporation (vaporization) in a vacuum state. However, when the liquid to be applied is an impure liquid, it is difficult to apply it to the desalination process because the packing surface is contaminated and performance degradation is expected. However, the vacuum bubble generation method is a phase change due to a local pressure drop that occurs in the middle of flow, and there is no source of problems such as contamination of the medium (thin film) or packing material.

Figure 2.46 is a conceptual diagram of a hypothetical desalination system using renewable energy sources and a vacuum bubbling mass evaporation device. This system is considered to be a desalination system for small-scale communities at the village level or smaller. The system consists of (1) vapor generator; (2) PV system, including hot water heater and electricity generation; (3) vapor condenser; and (4) vacuum pump and controller.

1. A vapor generator is a device that vaporizes water in a vacuum to produce vapor. The body of the vapor generator is a pressure vessel that can withstand vacuum, and a bubbler system (or a local heating method for

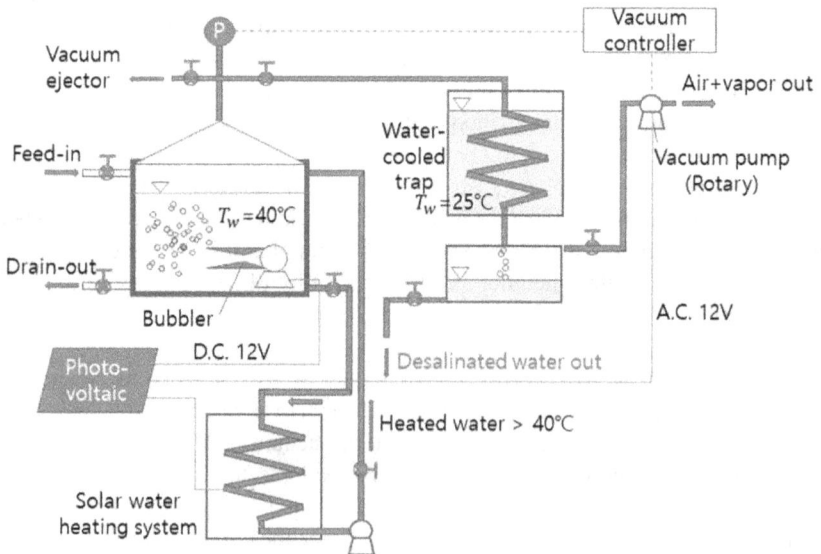

FIGURE 2.46 A conceptual diagram for vacuum bubbling desalination system combined with solar renewables.

evaporation may be applied, but in this book, the discussion will be limited to bubbling) is installed. Insulation can be applied to the outer wall of the container as needed, and it is preferable to install a visualization window for monitoring the operating conditions inside the container. An inlet and outlet for treated water and an inlet and outlet for a hot water heater are installed on the upper and lower parts of the container, and ports for gas discharge and cooling are installed on the upper part. The bubbler system installed inside the vapor generator is a combination of a watertight sealed water pump powered by external electricity (PV system in this case) and a venturi nozzle connected to the water pump. The inflow water is sent to the hot water heater before depressurization and bubble generation and is heated to a desired temperature. After evaporation, the concentrated treated water is discharged through the outlet.

2. The solar energy system, which supplies the electrical energy and thermal energy required for the operation of the desalination system, plays two main roles. One of them is to supply thermal energy to the solar water heater, which raises the temperature of the treated water as much as necessary to heat the water to the desired water temperature required by the vapor generator. As the temperature of the water rises, the saturated vapor pressure rises, so the vacuum level and bubbler installation conditions required for vacuum bubbling can be alleviated. Another role of the solar energy system is to supply the electrical energy required for operation, which the PV system will be responsible for. The electrical energy required in this system is a vacuum pump, a circulation pump, a water pump, and service electricity for the controller. Since the hot water heater piping is connected to the vacuum system, special care is required during operation, and the requirements for pressure resistance conditions need to be reviewed.

3. The vapor condenser is a device that produces fresh water by condensing the vapor produced by the vapor generator through cooling. The condenser can be considered a water-cooled jacket, and the condensed water (fresh water) is sent to the treated water (fresh water) storage tank at the bottom. A discharge port is installed at the bottom of the treated water storage tank, and a port and a pipe connected to a vacuum pump are installed at an empty space at the top.

4. The vacuum pump and controller play the role of creating and maintaining the level of vacuum necessary for the operation of the entire system and are characterized by being connected in series with the inside of the vapor generator and the vapor condensing device.

Let us now briefly explain how this system works. First, (1) the water to be treated is filled into the tank through the water inlet of the vapor generator. (2) The water is heated to a desired level (e.g., 40°C) via a solar water heater. (3) When the temperature of the water inside the vapor generator reaches a desired level, the pressure is reduced by a vacuum pump, and a large amount of vapor is generated through a bubbler system. When the pressure inside the tank rises due to vapor generation, the controller connected to the vacuum pump operates the vacuum pump to reduce the pressure to a desired level. This process is an intermittent process. The vapor escaping from the vapor generator is moved to the vapor condenser by the vacuum pump's pressure-reducing drive and the reduced pressure by condensation due to the low temperature of the vapor condenser and, in this process, is condensed through heat exchange with the water jacket. The condensed water (fresh water) is collected in the storage tank provided at the bottom of the condenser, thereby completing the production of fresh water. A special feature of this particular system lies in that degassing is performed first and then desalination follows. As we have already seen, vacuum bubbling can provide an optimized solution for degassing at room temperature. If the condenser is operated with a mixture of evaporated vapor and non-condensable gases, significant efficiency problems may occur, as pointed out by Wang et al. [32]. Therefore, high operating efficiency can be guaranteed by first completing deaeration and then linking the evaporation process through the same system. The benefits of preheating water in this system are clear. For example, if the water temperature is raised from an initial temperature of 25°C to 40°C, the saturated vapor pressure of water increases from 3.17 kPa to 7.39 kPa, which can be expected to reduce the level of the pressure required for bubble generation or to produce massive vapor. The special feature of the desalination system using vacuum bubbling is that although it does not eliminate the energy required for the phase-change energy of water itself, the energy source required for the process can be supplied through renewable energy, which is relatively low-level energy rather than fossil fuels.

REFERENCES

[1] C. E. Brennen, Cavitation and Bubble Dynamics. Oxford University Press, New York, NY, 1995.

[2] J. A. Hong, J. S. Lee, Y. D. Jun, "Degassing dissolved oxygen through bubbles under a vacuum condition," Proceedings of 7th Thermal and Fluids Engineering Conference (TFEC) Held at UNLV, Las Vegas, NV, April 2022. Hosted by American Society of Thermal and Fluids Engineers, pp. 1021–1033, TFEC-2022-41702.

[3] "Solubility," Wikipedia. https://en.wikipedia.org/wiki/Solubility

[4] IUPAC, Compendium of Chemical Terminology, 2nd ed. (the "Gold Book"), 1997. Online corrected version: (2006–) "Solubility." http://doi.org/10.1351/goldbook.S05740

[5] C. Tomlinson, "On supersaturated saline solutions," Philosophical Transactions of the Royal Society of London, 1868, 158, pp. 659–673. http://doi.org/10.1098/rstl.1868.0028.

[6] The Engineering ToolBox, "Solubility of air in water," 2004. www.engineeringtoolbox.com/air-solubility-water-d_639.html. Retrieved August 25, 2023.

[7] S. H. Yoo (ed.), Encyclopedia of Soil Science. Seoul National University Press, Seoul, 2000, pp. 443–444. ISBN 89-521-0204-5. (In Korean)

[8] NOAA (National Oceanic and Atmospheric Administration), The Atmosphere—Introduction to the Atmosphere. U.S. Department of Commerce. https://noaa.gov

[9] "Solubility – quantification of solubility," Wikipedia. https://en.wikipedia.org/wiki/Solubility

[10] D. W. Ball, Physical Chemistry, 2nd ed. (Korean). Sciplus, Seoul, 2017, p. 224.

[11] "Henry's law," Wikipedia. https://en.wikipedia.org/wiki/Henry%27s_law

[12] R. Sander, W. E. Acree, A. De Visscher, S. E. Schwartz, T. J. Wallington, "Henry's law constants (IUPAC Recommendations 2021)," Pure and Applied Chemistry, 2022, 94, pp. 71–85. http://doi.org/10.1515/pac-2020-0302

[13] R. Sander, "Compilation of Henry's law constants (version 4.0) for water as solvent," Atmospheric Chemistry and Physics, 2015, 15, pp. 4399–4981. http://doi.org/10.5194/acp-15-4399-2015

[14] R. Fernández-Prini, J. L. Alvarez, A. H. Harvey, "Henry's constants and vapor-liquid distribution constants for gaseous solutes in H_2O and D_2O at high temperatures," Journal of Physical and Chemical Reference Data, 2003, 32(2), pp. 903–916.

[15] A. H. Harvey, "Semiempirical correlation for Henry's constants over large temperature ranges," AIChE Journal, 1996, 42(5), pp. 1491–1494. https://doi.org/10.1002/aic.690420531

[16] International Association for the Properties of Water and Steam (IAPWS), "Release on values of temperature, pressure and density of ordinary and heavy water substances at their respective critical points," Physical Chemistry of Aqueous Systems: Meeting the Needs of Industry (Proceedings of the 12th International Conference on the Properties of Water and Steam), edited by H. J. White, Jr., J. V. Sengers, D. B. Neumann, J. C. Bellows. Begell House, New York, 1995, p. A101.

[17] P. M. Doran, Bioprocess Engineering Principle. Academic Press, London, 1995, pp. 206–207.

[18] A. Wüest, N. H. Brooks, D. Imboden, "Bubble plume modeling for lake restoration," Water Resources Research, 1992, 28(12), pp. 3235–3250.

[19] D. F. McGinnis, J. C. Little, "Predicting diffused-bubble oxygen transfer rate using the discrete-bubble model," Water Research, 2002, 36, pp. 4627–4635.

[20] A. Prosperetti, "Vapor bubbles," Annual Review of Fluid Mechanics, 2017, 49, pp. 221–248. http://doi.org/10.1146/annurev-fluid-010816-060221

[21] J. A. Hong, J. S. Lee, Y. D. Jun, "Extraction of oxygen-enriched-air from water through vapor bubble diffusion," Journal of Integrated Science and Technology, 2021, 9(1), pp. 22–29.

[22] ASTM D2779-92, Standard Test Method for Estimation of Solubility of Gases in Petrolium Liquids (Reapproved), 2012.

[23] Y. A. Cengel, M. A. Boles, Thermodynamics—An Engineering Approach, 5th ed. in SI Units. McGraw-Hill, New York, NY, 2006, p. 909.

[24] M. Frank, D. Drikakis, "Inert state of fuel tank during aircraft Ascent," AIAA SciTech Forum, 55th AIAA Aerospace Sciences Meeting, Grapevine, TX, January 9–13, 2017. Glasgow: University of Strathclyde.

[25] L. Cheng, L. Hua, J.-L. Yang, K.-X. Liu, "Simulation and analysis of crashworthiness of fuel tank for helicopters," Chinese Journal of Aeronautics, 2007, 20(3), pp. 230–235.

[26] National Transportation Safety Board, "In-flight breakup over the Atlantic Ocean Trans World Airlines Flight 800, Boeing 747–131, N93119 near East Moriches, New York July 17, 1996," Aircraft Accident Report NTSB/AAR-00/03, 2000. https://apps.dtic.mil/sti/pdfs/ADA388166.pdf

[27] S. Kinnersley, A. Roelen, "The contribution of design to accidents," Safety Science, 2007, 45(1), pp. 31–60.

[28] S. M. Summer, "Mass loading effects on fuel vapor concentrations in an aircraft fuel tank ullage," DOT/FAA/AR-TN99/65, 1999. https://www.fire.tc.faa.gov/pdf/tn99-65.pdf

[29] M. Burns, W. M. Cavage, "Inerting of a vented aircraft fuel tank test article with nitrogen-enriched air," DOT/FAA/AR-01/6, 2001. https://www.fire.tc.faa.gov/pdf/01-6.pdf.

[30] R. W. Fox, A. T. McDonald, Introduction to Fluid Mechanics, 4th ed. Wiley, New York, NY, 1994, pp. 336–341.

[31] S. Klein, Student Resource DVD Containing Engineering Equation Solver (EES) by F-Chart Software, 7th ed. in SI Units. Accompanying Thermodynamics an Engineering Approach, 6th ed. McGraw-Hill, New York, NY, 2008.

[32] L. Wang, X. Ma, H. Kong, R. Jin, H. Zheng, "Investigation of a low-pressure flash evaporation desalination system powered by ocean thermal energy," Applied Thermal Engineering, 2022, 212, p. 118523. https://doi.org/10.1016/j.applthermaleng.2022.118523.

3 Industrial Applications

The industrial applications of vacuum bubbling introduced so far have not yet been realized. Of course, degassing under a pressure higher than the saturated vapor pressure of the liquid has been used for a long time, but it is difficult to find reported case of industrial application of bubbling under a pressure lower than the saturated vapor pressure. Therefore, the industrial applications introduced in this chapter should be considered as prospective applications that the author personally thinks of, and these are only a limited subset of the numerous desirable applications this technology could find in the future.

Among the fields where vacuum bubbling may possibly be applied, this book introduces the following four fields: process degassing, "artificial gills" or underwater breathing, fuel degassing, and desalination. Each of these four areas has unique characteristics. Process degassing is a field that requires a shift from the current high-energy process to a low-energy-use process, and vacuum bubbling can be considered as an alternative technology that can reduce energy costs. In this field, where different processes require different levels of deaeration, vacuum bubbling appears to be able to provide the highest requirements for deaeration. Another interesting application of vacuum bubbling can be the use of highly oxygenated air, which is the result of degassing. Coincidentally, this comes from the fact that oxygen has a solubility that is about twice that of nitrogen in most solvents, including water. In the case of water, though the amount is limited, the gas obtained through degassing has higher oxygen content of about 50% than that of the atmosphere, so vacuum bubbling can be used for scuba divers or when underwater living space is under consideration. In addition, the role of degassing can be of high importance in aviation with fuels that are currently used and to be used in the future, such as sustainable aviation fuel (SAF). Pre-degassing of aviation fuel is believed to contribute not only to ensuring the safety of aircraft but also to controlling the oxidation of biofuels in general. The last possible application area but of great potential lies in the course of seeking solutions to global water shortage issues, which is the low-cost renewable energy–based vacuum bubbling desalination. The principle of generating vapor bubbles under a vacuum is the same phenomenon as vaporizing water by boiling it at high temperatures. Research results so far have focused on the fact that vapor bubbles can be created by using low-pressure conditions without resorting to heating. However, just as the nucleate bubbles appear in water at first and boils up when the conditions are ripe, vacuum bubbling should also be like that. Can we not hope that it can be achieved? Large-scale evaporation through vacuum bubbling may provide clues to solving the core energy demand of desalination with non-fossil fuel, low-level energy. Now, let us learn more about each field.

DOI: 10.1201/9781003374626-3

3.1   PROCESS DEAERATION

3.1.1   THERMAL DEAERATORS

Process degassing in industry is one of the widely applied engineering practices. In particular, the heated pressure degassing process, which is most commonly used in energy applications where heating of feedwater is routine (Figure 3.1), is a complex and large-scale facility that uses considerable energy while performing deaeration through a condition-sensitive process. The feedwater undergoes heating, evaporation, and condensation before finally being stripped by the steam supplied from a separate boiler. In this process, spraying or tray structures are used for better phase separation. For this reason, the traditional degassing process is known for its high energy costs, large initial equipment costs, complex structure, and difficult operation. However, thermal or pressure deaerator is still a widely used technology, as it is known to achieve the lowest dissolved oxygen concentrations in large quantities. Membrane deaerators have begun to spread as an alternative to pressure deaerators and are currently being applied to many processes. Conventional methods almost always tend to rely on high temperatures and high pressures. Recently, interest in low-energy processes has increased along with process efficiency improvement to curb the use of energy, especially from fossil fuels, and accordingly, alternative technologies using vacuum technology are being introduced one after another. Vacuum bubbling uses the same two principles used in the pressure degassing process, that is, solubility control and mass diffusion by vapor bubbles, but the operating conditions are different. In addition, you can expect deaeration performance comparable to that of pressure deaerators. Until now, laboratory research has been conducted to understand the underlying principles and explore the possibility of practical use, but it will be a matter of time and opportunity to apply it to various degassing fields in the near future.

For a general understanding of the technology and industry in the field of degassing, let us take a brief look at the contents in the literature. According to the Wikipedia dictionary, a *deaerator* is a device that removes oxygen and other dissolved gases from liquids and pumpable compounds [1]. Manufacture of deaerators began in the 1800s, with the US Navy introducing atomizing deaerators in 1934. Feedwater header and deaerator designs were improved in the 1920s and continue to be used in many applications today. This is the brief story of the development of thermal deaerators. A patent for a method for manufacturing deaerated water was endowed to George M. Kleucker in 1899, and a patent for degassing/degassing by steam bubbling in liquid, which is a key process of thermal degassing, was endowed to George Gibson, Percy Lyon, and Victor Rohkin of Cochrane from 1921 to 1933. The summary of one of these, "APPARATUS FOR TREATING LIQUIDS," registered on June 13, 1933, is as follows: [2]

> *In one type of such apparatus, the water is heated and deaerated by bringing it into intimate contact with flowing steam in a closed chamber. The water to be treated is passed through the chamber in a finely divided form and the steam coming into contact therewith heats the water to or substantially to the temperature of the steam,*

whereby air and other undesirable gases entrained in the water are separated there-from. Those skilled in this art understand that while oxygen and other permanent gases are soluble in water at all temperatures below boiling point, at the boiling point corresponding to any given pressure to which the water may be subjected, the solubility becomes zero, and on agitation of the water, the gases readily separate therefrom. The uncondensed portion of the heating and deaerating steam and the air and other gases liberated from the water are vented from the chamber. The heated and deaerated water is then usually passed into a storage tank adjacent the deaerating heater and withdrawn therefrom as required by pumping apparatus associated with the apparatus in which the heated used.

The technical details of this patent show the basic operating principle of the thermal deaerator currently in operation. So the issue of this topic goes back almost 100 years. Here, the spraying of water for degassing is referred to as "finely divided form," and the effect of heating and evaporation by steam and "agitation" is also explained. Stickle [3], Cochrane, and Permutit were the three oldest deaerator manufacturers in the United States; however, after a court case between the Elliott Company (no longer in business) and HSBW Cochrane Corporation in 1929, the two companies were allowed to continue manufacturing deaerators. After this, manufacturers Cochrane, Darby, Elliott, Gröeschel, Stearns-Rogers, Worthington, and others competed with each other for business, and in 1949, Chicago Heater was founded and became a leading deaerator manufacturer. In 1954, manufacturers Allis-Chalmers, Chicago Heater, Cochrane, Elliott, Graver, Swartwout, and Worthington went into business [4]. Wow! This thermal deaerator has been applied to the industry as a unique technology until the more energy-efficient membrane-type deaerator that has recently been introduced is popularized. Deaerators are used in many different industries, such as cogeneration plants, hospital boiler rooms, large laundry facilities, marine, oil fields (steam injection, etc.), oil refineries, offshore platforms, paper mills, power plants, prison boiler rooms, and steel mills.

Types of deaerator include thermal deaerator, vacuum deaerator [5], and ultrasonic deaerator, among which thermal deaerator includes spray type and tray type (also called cascade type). Thermal deaerators are commonly used to remove dissolved gases from feedwater for steam-generating boilers. Dissolved oxygen in feedwater can cause severe corrosion damage to boilers by adhering to the walls of metal pipes and other equipment and forming rust-like oxides. Dissolved carbon dioxide combines with water to form carbonic acid, which can cause further corrosion. Most deaerators are designed to remove oxygen to levels below 5–7 ppb by weight and essentially remove carbon dioxide. The deaerators in the steam generation system of most thermal power plants use low-pressure steam obtained from the extraction point of the steam turbine system. However, steam generators in many large industrial facilities, such as petroleum refineries, can use available low-pressure steam. There are many different deaerators available from different manufacturers, and the actual construction details will vary from manufacturer to manufacturer. Figure 3.1 shows an installed view of a typical thermal deaerator [6], and Figures 3.2 and 3.3 are representative schematic diagrams representing the two main types of degassing, respectively.

Figure 3.1   Installation view of a typical thermal deaerator—Courtesy Ignatus [6].

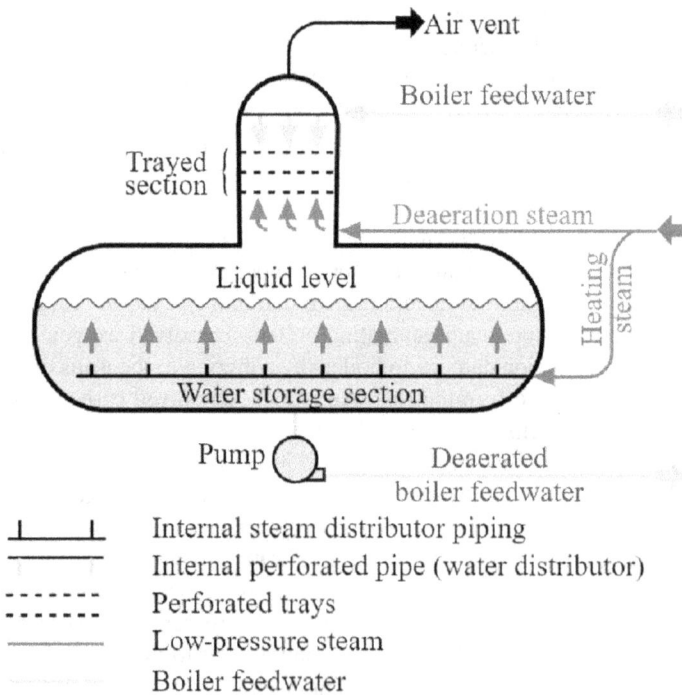

Figure 3.2   A schematic diagram of a typical tray-type deaerator [7].

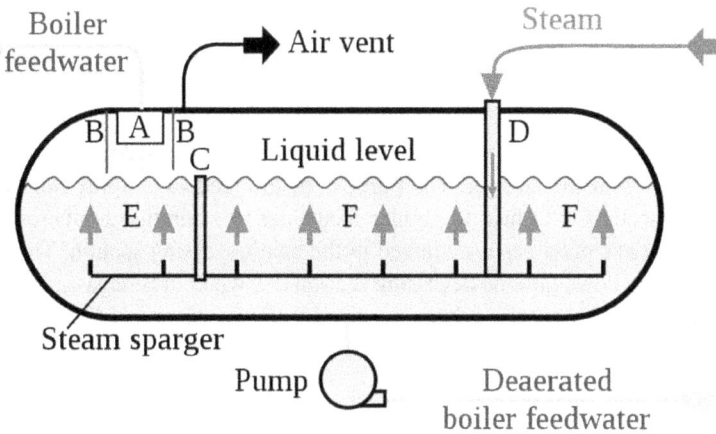

A = Spray nozzle
B = Spray nozzle shroud
C = Baffle
D = Steam supply pipe
E = Preheating section
F = Deaeration section

Figure 3.3 A schematic diagram of a typical spray-type deaerator [8].

Spray- and Tray-Type Deaerator

A typical tray deaerator, shown in Figure 3.2, has a vertical domed deaeration section mounted above a horizontal boiler feedwater storage vessel. Boiler feedwater enters the vertical degassing section through a spray valve above the perforation tray and then flows down through the perforations. The low-pressure stripping steam enters under the perforated tray and flows upward through the perforations. The combined action of the spray valve and tray ensures very high performance due to the long contact time between steam and water. The steam removes dissolved gases from the boiler feedwater and exits through a vent valve at the top of the domed section. The degassed water flows into a horizontal storage vessel, where it is pumped to a steam-generating boiler system. Low-pressure heating steam enters the horizontal vessel through a sparge pipe at the bottom of the vessel to keep the stored boiler feedwater warm. External insulation of the vessel is usually provided to minimize heat loss [1]. This is a very helpful article from Wikipedia's "Deaerator." However, the interesting part is the role of the sparger by low-pressure heating steam. Although the article describes its use to keep stored boiler feedwater warm, the more important role of steam may be the stripping of dissolved gases by steam bubbles. This is more clearly described in the spray deaerator introduced next.

Spray-Type Deaerator

As shown in Figure 3.3, a typical spray deaerator is a horizontal vessel with a pre-heating section (E) and a deaeration section (F). The two sections are separated by a baffle (C). Low-pressure steam enters the vessel through a sparger in the bottom of the vessel. Boiler feedwater is injected into section (E), where it is preheated by steam rising from the sparger. The purpose of the feedwater spray nozzle (A) and preheating section is to heat the boiler feedwater to saturation temperature so that dissolved gases can be easily removed in the next degassing section. The preheated feedwater then flows into the degassing section (F), where it is degassed by the rising steam in the sparger system. Gases removed from the water are discharged through the vents at the top of the vessel.

Vacuum (Dynamic) Deaerator

Wikipedia also introduces the vacuum deaerator and its special form, the dynamic deaerator (Figure 3.4 [9]). As can be seen in the figure, the liquid to be treated in the dynamic deaerator is distributed in a thin layer on a high-speed rotating disk via a special feed system. The centrifugal force throws the sling through a perforated screen and onto the inner wall of the vacuum vessel. Air (gas) pockets are released from the process and evacuated by vacuum. This type of deaerator is also used to remove dissolved gases from products such as food, personal care products, cosmetics, chemicals, and pharmaceuticals to increase dosing accuracy in the filling process, increase product storage stability, and prevent oxidation effects (e.g., discoloration, change in smell or taste, rancidity). Vacuum deaerators are also used in the petrochemical sector.

3.1.2 MEMBRANE DEAERATORS

Thermal deaerators, which have been active in this field for a long time, have disadvantages in energy efficiency as well as high initial costs due to large and complex facilities, as mentioned earlier. With the recent development of membrane materials, the development of degassing using membranes is accelerating, and the scope of application is also expanding. Membrane deaerator is a relatively recent technology, and commercial technology development is in full swing, and most of the sources of available data are related to product promotion. The level of technology in this field is examined through a few examples.

On the homepage of Canon Artes [10], one of the companies providing membrane degassing solutions, "[m]embrane deaeration is [described as] an ideal solution for the quantitative removal of oxygen and carbon dioxide at ambient pressure," while mentioning that the membrane degassing process is introduced as follows:

> In the membrane degassing process, water is brought into contact with the purge gas stream through a semi-permeable membrane. Water cannot permeate through the membrane, but dissolved gases are forced to flow through the membrane by the driving force of the low partial pressure of the gaseous state. Depending on each customer's expected

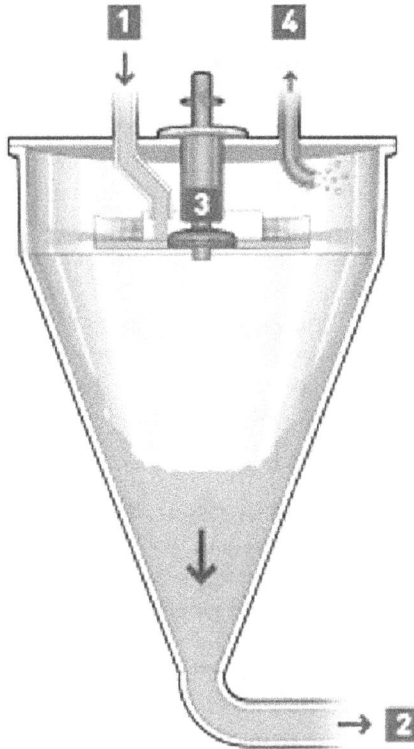

FIGURE 3.4 A schematic diagram of rotating disc deaerator [9].

performance, the purge gas stream can be pure nitrogen, compressed air, under vacuum, or a combination.

Now, let us take an overview of the prospects of degassing technology using membrane technology based on the presentation data on the characteristics of the membrane contactor by Paul Peterson of 3M at the Produced Water Society Seminar in 2016 [11]. He explained the principle of gas transfer in a membrane contactor in the following three steps: (1) Gas in the atmosphere dissolves in water until equilibrium is reached. (2) The equilibrium between liquid and gas is altered when vacuum and/ or sweep gas is applied. (3) This change in the equilibrium condition generates a driving force that moves the gases in the liquid into the gas phase. The membrane contactor can operate in four modes: (1) sweep gas mode, (2) vacuum mode, (3) mixing mode, and (4) blower mode for carbon dioxide removal. By changing the partial pressure, you can either remove the gas from the water or dissolve the gas into the water. The principle is that when the partial pressure on the gas side is increased, the gas dissolves in the water, and when the pressure is lowered, the gas is removed from the water.

Among the data presenting the degassing performance of this unit by operation mode, the theoretical minimum achievable DO level is known to be a function of applied vacuum level and temperature in the vacuum-only mode. The application range of vacuum varies depending on the operating temperature conditions (5°C to 60°C), which seems to be limited by the limitations of the physical properties of the membrane depending on the temperature. The degassing performance of a membrane contactor in vacuum-only mode is a function of temperature and applied vacuum pressure, and the theoretical minimum DO level achievable at a temperature of 45°C and a pressure of 72 mmHg (absolute pressure of about 9.6 kPa) is reported to be about 5 ppb.

The general advantages of membrane degassing can be derived from the manufacturer's claims regarding the advantages of membrane contactors; it is small in size, is compact, has low installation cost, has excellent expandability as it is in a modular unit, is easy to clean, can theoretically propose a low dissolved gas concentration, and does not require chemicals during operation. It is environmentally friendly and safe for workers.

However, in the case of membrane degassing, there is a problem of fouling due to biological or mineral substances that may be included in the treatment process. There is a restriction that replacement is required, and there is also a risk factor due to membrane failure. A good description of the industrial applications of membrane-based degassing is the brochure on 3M™ Liquid-Cel™ membrane contactors [12]. This material covers the main concepts and applications related to the removal, injection, or control of dissolved gas in liquid with a more expanded concept called gas transfer membrane technology. However, in this text, only the contents related to degassing are selectively introduced.

1. *Beverage industry.* In the beverage industry, degassing contributes to extending the shelf life of beer, wine, juice, and many other beverages.
2. *Laboratory water and analytical instruments.* Dissolved gases may interfere with instrument readings and may alter the composition of the analytical solution. Effective degassing of laboratory water improves measurement reliability.
3. *Oil and gas production.* In the oil and gas industry, water is degassed to protect boilers and to remove environmental pollutants such as ammonia.
4. *Pharmaceutical manufacturing.* The pharmaceutical industry relies on low CO_2 levels in process water to control water conductivity and pH, so CO2 is degassed to maximize the efficiency of the electrodeionization process. It also removes O_2, reducing the risk of oxidation and spoilage of the final product without the use of chemicals.
5. *Power and steam generation.* By removing dissolved O_2 and CO_2 from makeup water, the risk of corrosion and pitting due to oxidation and carbonation is reduced. This helps power plants reduce the high cost of capital equipment maintenance and replacement by reducing long-term damage to pipes, fittings, and boiler surfaces.
6. *Printing, inks, and coatings.* Widely used to remove dissolved gases from inks and coatings so that production lines work efficiently and printheads eject ink smoothly and precisely.

7. *Semiconductor and microelectronics.* For semiconductor processing, flat panel display manufacturing, and other high-tech industries, water degassing technology is applied. Today's plant requirements for CO_2 and O_2 in water are extremely low. Low O_2 levels protect products, and low CO_2 levels help maximize the efficiency of water electroionization and ion exchange processes.

3.1.3 VACUUM DEAERATORS

So far, we have briefly looked at traditional thermal degassing and degassing using membranes, but the recent climate catastrophe reveals the fact that curbing fossil energy consumption, supplying renewable energy, and improving energy efficiency are the priorities of all mankind [13, 14]. In the field of degassing, much attention is being paid to degassing methods that consume less energy. Most of the recently introduced technologies (products) are vacuum-based technologies. Although the vacuum deaerator was introduced earlier, it would be interesting to see what technologies are being applied through the vacuum deaerator products ranked at the top of Google search (Table 3.1).

Most of the published information on the vacuum degassing method are promotional materials related to commercialization, and there is not much detailed open technical information. Based on limited published data, it seems that vacuum degassing technology lowers the phase-change temperature condition by using vacuum in the thermal degassing method and makes evaporation easier through surface modification. Here, thermal degassing focuses on gas separation through the phase-change process of heating-evaporation-condensation, and vacuum degassing in this sense can be seen as low-pressure (vacuum) thermal degassing. The second characteristic of vacuum degassing is in that they maximize the surface area between phases and to create conditions for dissolved gases to easily escape by providing dynamic conditions. In the case of dynamic degassing without the phase-change process of heating-evaporation-condensation, even if heating is involved, degassing levels seem limited by the solubility.

According to [25], vacuum deaerators first reduce the partial pressure of oxygen in the gas phase to mechanically remove dissolved oxygen from water and then accelerate the transfer of oxygen from liquid to gas phase by using a uniquely shaped packing as a contacting device. The expected feature of this device is to maximize surface area while minimizing liquid hold-up and pressure loss.

In addition, Eurowater's homepage provides useful information by introducing water treatment processes, including degassing, in various industries, including green hydrogen production that requires pure water (www.eurowater.com/en/applications) [26].

Compared to the other deaerators mentioned earlier, recently introduced corosys's V2WD [27] seems to show relatively superior performance in dissolved oxygen concentration requirements of less than 10 ppb. Most vacuum degassing seems to involve some degree of heating, and in the case of corosys' CWD model without heating, the dissolved oxygen concentration requirement is relaxed to 50 ppb, which can be found on the website [24]. The system is reported to be characterized by high operational reliability and low energy consumption. From Table 3.1, in the case of

TABLE 3.1

Vacuum Deaerator Suppliers and Product Features

No.	Manufacturer (Source)	Type (Capacity)	Usage	Water Temp.	DO Level
1	Eurowater [15]	Filler 22 m³/h	District heating	40–90°C	0.2 mg/L
2	Eurowater [15]	Filler 1 m³/h	All-in-one	40–90°C	0.2 mg/L
3	LMTECH [16]	Vacuum, dynamic	Food, cosmetics, coating, bonds, paint, ink, chemicals	NA	NA
4	EWT Water Technology [17]	Thermal	District heating	45–85°C	20μg / L
5	Veolia (Whittier Vacuum Deaerator) [18]	Packed tower 100–10,000 gpm	Industrial-grade, mining	NA	50 ppb
6	AWC [19]	Filler	Heating plants	40–90°C	0.2mg/L
7	Cannon Artes (ZeroGas vacuum deaerator) [20]	Packed tower	District heating, off-shore water treatment	NA	NA
8	BACHILLER [21]	Dynamic	Pharmaceutical, cosmetics, food, chemical	NA	NA
9	PerMix [22]	Dynamic 1,500–30,000 lpm	Pharmaceutical, cosmetics, food, chemical	NA	NA
10	PROMACH-TECHNIBLEND (TB-D Series) [23]	Spray-vacuum	Beverage	NA	NA
11	Corosys beverage technology-V2WD [24]	Packing 5–100 m³/h	Brewing, beverage, food, chemical, pharmaceutical, power	NA	<10 ppb

vacuum degassing, the final dissolved oxygen concentration that can be provided differs depending on the deaeration method applied and the fields of application. It can be seen that the presented data are somewhat relaxed compared to the requirement of 5 ppb that thermal deaerator boasts. Why? This may be related to the lower solubility limit due to heating—solubility tends to decrease at higher temperatures and lower pressures—but the most important difference may lie in the degree of steam involvement. Though not clearly stated in the description of the presented vacuum degassing systems, the feasibility of achieving the minimum dissolved oxygen concentration level of 5 ppb or less seems to depend on whether scavenging by vapor or steam bubbles is performed or not. Looking back at the available information on the degassing methods so far, it could be seen that innovative technologies that are simpler and smaller in structure, consume less energy, are environmentally friendly, have improved performance, and will continue to emerge.

3.1.4 VACUUM BUBBLING DEAERATION—A CASE STUDY

Vacuum bubbling is still in its infancy, but based on the results of research so far, it is expected to be a practical technology that fits well with the technology development trends of needs.

A dissolved oxygen removal device for proton accelerator RCCS (resonance controlled cooling system) is considered as a simple example. Basic requirements for the design are given as follows:

- Flow rate: 1.8 m³/h (30 lpm)
- Water temperature: 19°C to 30°C during operation
- DO concentration requirement: not to be opened
- Size limit: $L \times W \times H \leq 0.33\,m \times 0.45\,m \times 1.3\,m$
- Material: stainless steel

The conceptual design of the system for this application is shown in Figure 3.5. The proposed container is a rectangular stainless steel (STS 304) with a size of $L \times W \times H \leq 0.3\,m \times 0.4\,m \times 0.6\,m$, but the upper cover can be opened as needed, and

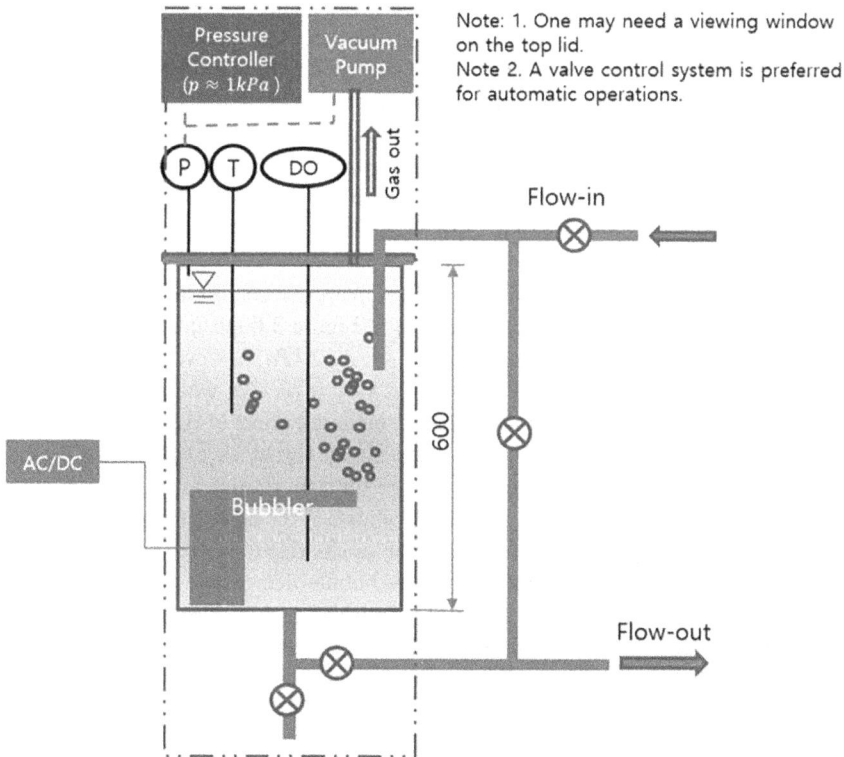

FIGURE 3.5 A conceptual design of vacuum bubbling deaerator (a duplicated figure of Figure 1.8).

various sensors (temperature, pressure, and dissolved oxygen), power lines, and treated water inlets and gas vent ports are to be installed. Also preferably, a visualization window is installed on the upper lid. Inside the container, a bubble generator driven by a DC power source is preferably fixed in a horizontal direction at a height of about 20 cm from the bottom surface. The bubble generator is a combination of a water pump and a venturi nozzle, and preferably, the nozzle is installed to keep it leveled. In the case of water pump, sealing under repeated operating conditions would have to be guaranteed in advance. The bubble generator can use DC power (DC 12V) supplied from outside of the tank. Water to be treated is introduced through the water inlet port installed on the top of the container. After the degassing, it can be sent either to the return water line (flow-out) or recycled as needed. An outlet is provided in the drain passage from the tank together with the water line. The vacuum pump discharges the air inside the container through the vent port on the top of the container to lower the pressure inside the container to the desired level. In order to prevent contamination of the vacuum pump, a separate moisture removal trap may be applied as necessary. The pressure controller preferably operates in conjunction with the pressure gauge to maintain the pressure inside the vessel within a desired range. In this design, temperature control such as separate heating is not applied.

Working principle and concept. Treated water flows into the vessel through the inlet, and the water is filled to a height of about 50 cm. In this case, the expected depth of the nozzle is 0.3 m from the water surface. Considering the temperature range (19°C–30°C), the representative temperature of water is assumed to be 25°C for convenience. (The saturated vapor pressure of water is $p_{vap} = 3.17\ kPa$.) Maintain the decompression level at $p_1 = 1\ kPa$ and start degassing by operating the bubbler. The operating condition, as illustrated in Figure 3.6, corresponds to $p_4 = 3.93\ kPa$, and the same bubbler unit and input power as in the basic experiment in Sec. 2.1.3 may be used. It is possible to predict the degassing performance based on the experimental results discussed in Sec. 2.3.2.4.

Analysis of the operating conditions. The operating conditions of the proposed system on the phase-change diagram shown in Figure 3.6 tell that the tank internal pressure $p_1 = 1\ kPa$ is already lower than $p_{vap} = 3.17\ kPa$; however, the vapor retention condition becomes $p_4 - p_{sat} = 3.93 - 3.17 = 0.76\ kPa$, which represents mild adverse condition for vapor retention, due to the depth effect of 0.3 m from the water surface. Therefore, the bubbles formed at the nozzle throat are exposed to a pressure higher than the saturated vapor pressure, albeit for a short time, at the downstream of the nozzle, so some bubbles may shrink or collapse at the downstream of the bubbler nozzle. During this process, the survived bubbles rise due to buoyancy. As the bubble rises, the hydrodynamic pressure around the bubble decreases, so after the bubble passes the depth of 0.22 m (the depth at which the pressure becomes equal to the saturated vapor pressure), the size of the bubble rather starts growing.

Now, let us predict the performance of the deaeration model shown in Figure 3.5 using the deaeration model introduced in Sec. 2.2.2. First of all, the basic data needed to predict performance include:

- Volume of water: 0.060 m^3 (= 0.3 m × 0.4 m × 0.5 m) with the assumed water depth of 50 *cm*

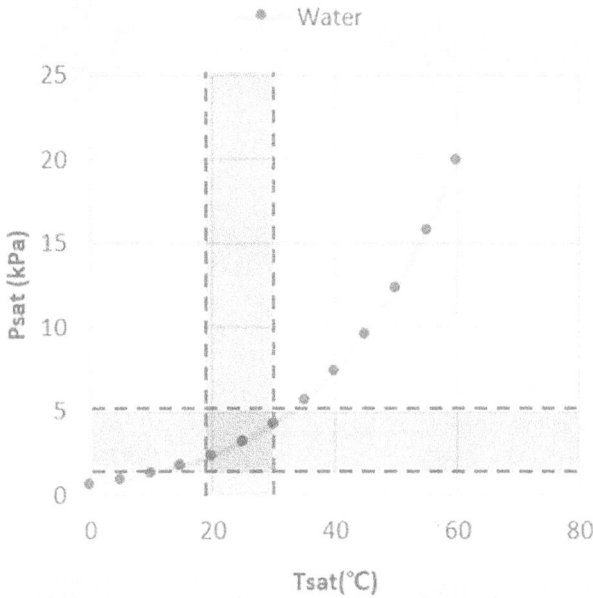

FIGURE 3.6   Expected operating conditions of the proposed deaeration system.

- Operating temperature: 19°C–30°C
- Degassing rate: correlated data from experiments (described in Sec. 2.3.2.4)

Figure 3.7 is the correlation data between dissolved oxygen concentration and degassing rate, which was obtained based on the experimental results when the water temperature was 19°C, which is the experimental condition described in Sec. 2.3.2.4. It is obtained from regression equations for the dissolved oxygen concentration and degassing amount (rate) with respect to the bubbling elapsed time, and by considering the amount of bubbles constantly generated at the later stage of the degassing process.

$$Q_N(t) = 0.0042 \times C_{DO}^{1.1574} \text{ for } C_{DO} \geq 0.2 \, mg \, / \, L$$

$$Q_N(t) = 0.00073 \text{ for } 0 < C_{DO} < 0.2 \, mg \, / \, L.$$

Based on the preceding data, the change over time of the dissolved oxygen concentration at the same water temperature is shown in Figure 3.8. According to the predicted results, it was predicted that it would take about 15 min to reach the dissolved oxygen concentration of 1 mg/L and about 220 min to reach 0.1 mg/L. The present simulation results provide the performance behavior of the proposed deaerator based on the degassing rate information from the experiments.

Summarizing the aforementioned performance prediction results, when using the bubbler system used in the experimental device in Chapter 2, it takes about 220 min

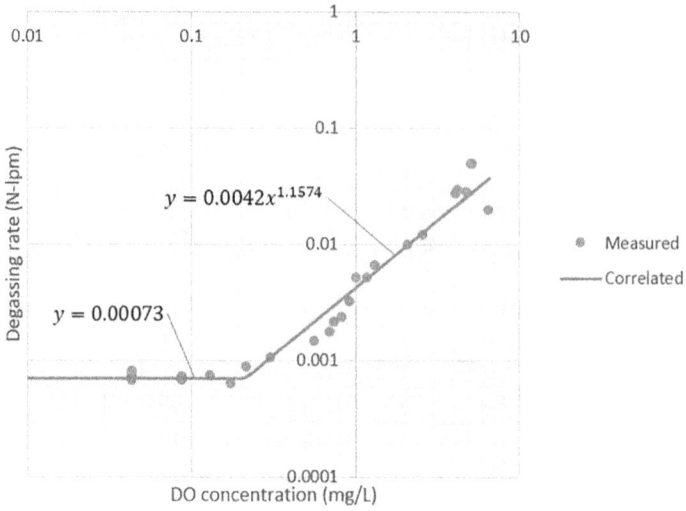

FIGURE 3.7 Correlation between DO concentration and degassing rate based on the measured data.

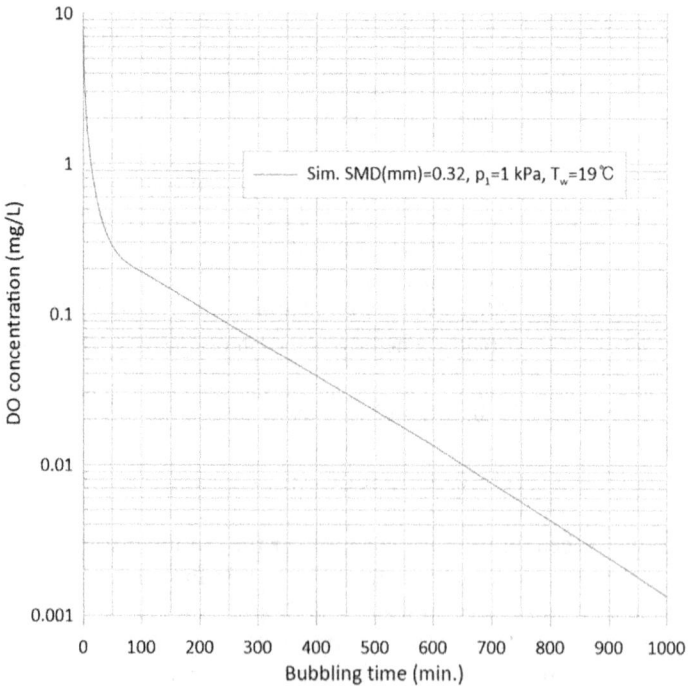

FIGURE 3.8 Estimated DO concentration for the test case (Water volume: 60 L, Bubbler input power: 20W).

to degas 60 L of water at room temperature to a level of 0.1 mg/L without heating, and the amount of electrical energy used is the sum of 20 W × 220 / 60 hr = 0.073 kWh (for driving the pump) and 0.5 kW × 220 / 60 hr × 0.3=0.55 kWh (for driving the vacuum pump), which is 0.623 kWh. Here, the operating time of the vacuum pump is based on a rough assumption of 30% of the total bubbling time based on the previous experimental experience.

Vacuum bubbling has the advantage of being able to achieve a high level of degassing with a very simple structure when applied to industrial degassing, and in particular, the energy cost required for operation can be expected to be reduced by an order of magnitude compared to existing processes. So far, the results of degassing experiments with no heating have been introduced, but it is expected that optimal operating conditions, including moderate heating, can be found by combining the four performance variables discussed in Chapter 2. Vacuum bubbling is also more environmentally friendly because it does not use intermediate materials such as membranes. Therefore, vacuum bubbling degassing is expected to be applicable from industries where energy efficiency improvement and eco-friendly issues are important among the spectrum of degassing devices across all industries. So is this approach perfect? Of course not. So far, as the author understands, the technology has not yet been optimized, so it will be necessary to establish optimal operating conditions for each application. Regarding the fabrication point of view, good sealing of pumps that operate in a vacuum state needs to be guaranteed all the time. Above all, scale-up technology from small to large devices also remains a challenge as applications become larger. However, solving these problems seems only a matter of time, and it is expected that in the near future, advanced industrial technology will be linked to vacuum bubbling, and energy-saving degassing is expected to become a routine in many industries.

3.2 FUEL DEGASSING

Although the removal of dissolved gases in fuel, particularly dissolved oxygen, is a necessary measure to ensure the safety of aircraft passengers, it seems that no engineering solution has been implemented to properly and safely deal with this problem. Instead, all airplanes, without exception, are supposed to be equipped with some kind of fuel tank inerting system on board. The dissolved air in the fuel is expelled from the fuel according to Henry's law when the surrounding pressure is lowered. If an airplane flies at a high altitude of about 10,000 m (or about 32,800 feet), the atmospheric pressure at that height is 26.5 kPa, which is about 1/4 of the standard atmospheric pressure, so about 3/4 of the dissolved gas at ground level is destined to be released. In the case of aircraft fuel, as in water, the solubility of oxygen is higher than that of nitrogen, so once the dissolved air escapes, it has a higher oxygen concentration than the atmosphere. Because this oxygen-rich gas can pose a safety concern if left in the tank ullage, the FAA controls oxygen levels inside the fuel tank ullage of an aircraft to ensure that they are below standard under all flight conditions. The interest in deaeration of aircraft fuel by vacuum bubbling began with the idea that it would be a more desirable method if deaeration was possible before the fuel was pumped into the fuel tank, rather than treating after thereof. None of the processes so far have been considered reliable for degassing dissolved gases in air-craft fuel, but there is room for application of deaeration by vacuum bubbling at the final stage of the production process or just before being pumped from storage tanks or tank lorries to aircraft fuel tanks, which will ensure that passengers will have safer flight conditions. Although a naïve idea, it is an appeal to the FAA to consider this approach for safer travel for air passengers around the world.

Another demand for degassing applications for aircraft fuels is related to biofuels. At a conference recently held in the United States, a professor from a national university in Brazil introduced her study on the development and application of biofuel. The interesting part was that the biofuel production plant should be located near the place of use due to the oxidation problem of fuel. Her argument was that because biofuels are easily oxidized, they easily deteriorate shortly after production, which is a major constraint on commercialization. This issue was intriguing, so I did some research and found that it is a common issue with biofuels, including sustainable aviation fuel (SAF). Current approaches in this area have almost always been additives. However, if the core of the problem is oxidation, would it not be helpful to remove the cause of oxidation in advance by degassing the dissolved oxygen in the fuel? Another important advantage of vacuum bubbling can be highlighted here. This is because it will not be an easy task to confirm the applicability of fuel degassing by heating or membrane degassing, but vacuum bubbling may work fine for this.

3.2.1 FUEL DEGASSING

The need for degassing of fuel, especially aircraft fuel, is that the causes of large air-craft explosions are related to the condition of fuel and fuel tanks. A typical example is the TWA 800 crash accident in 1996. On July 17, 1996, a Boeing 747 en route to Paris exploded over the Atlantic Ocean off the coast of Long Island just minutes after taking off from New York's Kennedy International Airport, killing all 230 people on board. The plane was blown apart in a violent explosion 12 min after taking off at 8:19 p.m. Among the dead were 18 crew members and 212 passengers, including 16 students and five chaperones from the Montoursville, Pennsylvania, area high school French language club [28]. An investigation concluded that the cause of the crash was not a terrorist attack but an electrical failure that ignited the 25-year-old aircraft's nearly empty center wing fuel tank. The incident went down in history as one of the deadliest plane crashes in American history. Following a final report released after a four-year investigation, the National Transportation Safety Board issued several safety recommendations, including a routine maintenance program and fuel tank design standards, to help prevent further accidents.

The oxygen concentration in the fuel tank ullage is the link between the electrical failure (presumed to be a spark) and the fuel tank ignited, causing the explosion. So how does oxygen concentration relate to combustion? To explain this, the principle can be explained through a simple experiment in which a candle placed in an enclosed space burns and consumes oxygen in the space, eventually turning off. Could there be no more oxygen in the inner space where the candle is extinguished? It is not like that. As a result of my own experiments and measurements, an oxygen concentration of more than 10% was measured even in the air with the candles turned off. This means that there is a range of oxygen concentrations within which combustion can proceed, and once that limit is exceeded, further combustion will not proceed. In view of this point, it seems that modern aircraft design is presenting requirements for ullage oxygen concentration [29–32]. Approaches to inerting fuel tanks to ensure safety during mission performance of aircraft known so far include explosion-prevention foam, halogen fire extinguishing system, liquid nitrogen, and NEA (nitrogen-enriched air) supply method through OBIGGS (on-board inert gas generating system) [32, 33]. What these approaches all have in common, however, is that they are processed after the fuel is injected into the airplane's fuel tank.

Then, how much of the gas is dissolved in the fuel and can be released during flight [34]? According to the calculation criteria for gas solubility in petroleum liquids [35], the *solubility of a gas* is defined as the volume of dissolved gas per volume of liquid when the gas and liquid are in equilibrium at a specified gas partial pressure and a specified temperature, which is called the Ostwald coefficient. Ostwald coefficient values for JP-8 fuel read 0.218 and 0.110 [36] at a temperature of 20°C and partial pressures of oxygen and nitrogen of 1 atm, respectively. These values are close to twice that of nitrogen for oxygen, similar to water, and also, comparing these values with those of water [37–39] (0.033 and 0.0165 for oxygen and nitrogen, respectively), oxygen is 6.6 times (0.218/0.033), and nitrogen 6.7 times (0.110/0.0165), more soluble, which are significant.

I propose a relatively simple and inexpensive way to degas aircraft fuel before it is injected into aircraft fuel tanks through non-thermal vacuum bubbling. However, it seems that a more comprehensive review of this issue is needed, since all the other issues that can arise from such degassing have not been fully considered. Among the issues to be reviewed, the suitability of the change in the combustion performance of the fuel due to degassing, and whether or not the various requirements for the fuel are satisfied, will have to be reviewed. So far, there has been little opportunity to discuss these issues, probably because any available technologies to be reviewed have not been identified so far.

The "degassing tank lorry" shown in Figure 3.9 was an idea the author submitted to the *Create the Future Design Contest* hosted by SAE in the United States in 2022. It was a fact that I learned while traveling in the United States after the exhibition, but it seems that large hub airports in the United States no longer use tank lorries and have underground supply facilities. However, many small- and medium-sized airports, such as Jeju International Airport, one of Korea's popular tourist attractions, still rely on tank lorries. The concept of the proposed work is roughly like this. As shown in the figure, the tank lorry with deaeration function adds several additional elements to the existing tank lorry. They are a bubbler system linked to an external battery, a vacuum pump and vent passage, and sensors to monitor the internal conditions (temperature, pressure, and dissolved oxygen). If applied in practice, a visualization window or alternative device for visually monitoring the internal situation may be required. When fuel is supplied to the tank lorry from the large storage tank, depressurization is performed through a vacuum pump while the fuel is full. Since the level of deaeration for fuel is only to satisfy the inerting requirements of ullage, the level of decompression required may be relatively moderate compared to other

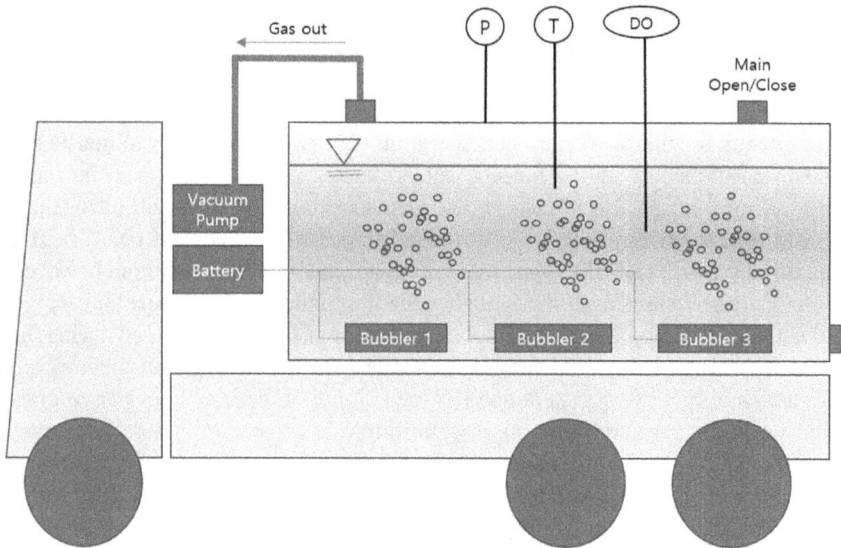

FIGURE 3.9   A conceptual view of "degassing tank lorry."

industrial degassing. In other words, the deaeration in this case seems to be sufficient with the first phase of vacuum bubbling as introduced in Sec. 1.9, in which the supersaturated solute is discharged by reduced pressure. As discussed in Sec. 2.3.2 and Sec. 3.1.4, this degassing process can be very fast. If there is a requirement for time, methods such as increasing the capacity of the bubbler pump or increasing the number of bubbler assemblies may be applied to accommodate it. After confirming that the desired level of degassing has been met within the specified time, the tank lorry's fuel is safely injected into the aircraft.

What are the technical limitations in realizing a degassing tank lorry? Among the important technical issues that need to be addressed may include sealing first and the compatibility of materials. In the case of sealing, it is a common problem to be solved even in water, but it seems that the problem of material compatibility (hardening and deterioration of plastic materials) needs to be solved. The application of jet fuel degassing using vacuum bubbling technology seems to be a task that can be easily realized with relatively little effort and cost. However, in order to advance this to practical use, a comprehensive technical review will be required under the intervention of related regulatory agencies. In addition, this issue is thought to be an application field that requires serious consideration and review because it may conflict with the business or economic interests of the industry that relies on current technology.

Now, let us take a quick look at biodiesel and its aviation version, sustainable aviation fuel (SAF). Biodiesel is known as a fuel that replaces or mixes petroleum-based diesel by making fuel similar to diesel through chemical processing of vegetable oil or animal fat [40–44]. Biodiesel was used earlier than petroleum diesel, as opposed to bioethanol later than gasoline. It is said that Rudolph Diesel used peanut oil as a diesel fuel he developed himself, but later, as diesel refined from petroleum was discovered, the existing diesel was named biodiesel. Biodiesel, like bioethanol, is named according to the mixing ratio. If the ratio of biodiesel to diesel mass is n%, it is expressed as Bn. For example, a 5% biodiesel blend is B5. In a search of the Internet related to biodiesel, many supply plans were announced mainly around 2010, but there have been fewer related articles recently. When I asked the Google Bard why, I got the following answer.

[Question 1] Are the plans of each country for the distribution of biodiesel still valid? What are the reasons, if any, for the plan to be delayed?

[Google Bard answer 1] Yes, the plans of each country for the distribution of biodiesel are still in effect. However, the reason for the delay in the plan is as follows.

- *Production cost*

  The problem of biodiesel's high production cost remains unresolved. The raw material for biodiesel is vegetable or animal fat, which is more expensive than petroleum. In addition, the production cost is high because the biodiesel production process is complex and energy intensive.

- *Raw material supply*

  The raw materials needed to produce biodiesel are vegetable oils or animal fats, which are limited resources. There are also concerns that using food crops for biodiesel production could lead to food shortages.

- *Technical issues*

  The production process of biodiesel has not yet been technologically perfected. By-products from the biodiesel production process can have adverse environmental impacts. Also, since biodiesel needs to be mixed with petroleum, it must maintain the same quality as petroleum. However, because biodiesel has a higher viscosity and higher acid value than petroleum, blending with petroleum can degrade the performance of the fuel.

  These problems are delaying the spread of biodiesel. However, since biodiesel is an eco-friendly fuel that can replace petroleum, research to solve these problems continues.

[Question 2] To what extent does the oxidation control issue of biodiesel have an impact on solving technical problems and commercializing it?

[Google bard answer 2] The issue of oxidation control of biodiesel has a great impact on solving technical problems and commercialization. Biodiesel is easily oxidized, and oxidation during storage and transportation can lead to reduced fuel quality and reduced engine performance. Therefore, controlling the oxidation of biodiesel is an essential technical issue for the commercialization of biodiesel.

Various methods are being studied to control the oxidation of biodiesel. The most common way is to add antioxidants. Antioxidants prevent oxidation of biodiesel by inhibiting oxidation reactions. In addition, methods for removing the cause of oxidation by improving the manufacturing process of biodiesel are being studied.

The oxidation control issue of biodiesel is an important issue to be resolved for the commercialization of biodiesel. Various methods are being studied for technical problem solving and commercialization, and it is expected that the commercialization of biodiesel will expand in the future.

[Question 3] Please introduce three international efforts or programs related to the development and distribution of biodiesel in order of importance.

[Google bard answer 3] International efforts or programs related to the development and distribution of biodiesel are as follows.

1. ***International Energy Agency (IEA)** is operating various programs for the development and distribution of biodiesel. The IEA supports biodiesel production technology development, biodiesel supply policy establishment, and biodiesel market research.*
2. ***The European Union (EU)** is implementing a biodiesel compulsory policy for the development and distribution of biodiesel. The EU requires a blend of at least 10% biodiesel in all diesel fuel by 2020.*
3. *The United States has implemented a biodiesel tax credit policy for the development and distribution of biodiesel. The United States is reducing the cost of biodiesel production by deducting taxes on companies that produce it.*

** represents (1) emphasis of importance and (2) formal designation, other than just using quotation marks.

These international efforts and programs are promoting the development and distribution of biodiesel, and it is expected that the use of biodiesel will expand in the future.

[Question 4] Please introduce the definition, development history, and vision of sustainable aviation fuel.

[Google bard answer 4]

Sustainable aviation fuel (SAF) is an eco-friendly fuel that can drastically reduce carbon emissions while maintaining the same performance as Jet-A, the existing aviation fuel. SAF is manufactured using renewable raw materials such as vegetable oil, animal fat, or waste, and can be blended with conventional aviation fuel.

Development of SAF began in the 1990s, and commercialization began in the early 2000s. Currently, SAFs are produced and used worldwide, and their production is increasing every year.

SAF can dramatically reduce carbon emissions compared to conventional aviation fuel. Because SAF is manufactured from renewable raw materials, its carbon footprint is very low. For example, SAFs can reduce carbon emissions by up to 80% compared to petroleum-based aviation fuels.

SAF is expected to play a key role in achieving carbon neutrality for the aviation industry. The aviation industry is one of the largest carbon emitters worldwide. SAF is expected to help combat climate change by reducing the aviation industry's carbon footprint.

Because SAFs are still expensive to produce, technology development and policy support are needed to mass-produce them. However, the potential of SAF is very high, and the production of SAF is expected to increase in the future.

The development of SAF is an important challenge for the aviation industry to achieve carbon neutrality. SAF can dramatically reduce carbon emissions compared to conventional aviation fuel, and is expected to play a key role in achieving carbon neutrality in the aviation industry.

[Question 5] Please introduce the difference between biodiesel and SAF and the development status.

[Google bard answer 5] Both biodiesel and SAF are eco-friendly fuels that can replace petroleum. However, there is a difference in that biodiesel uses vegetable oil or animal fat as a raw material, and SAF uses a renewable carbon raw material as a raw material.

Biodiesel was first developed in the 1970s and is now widely used worldwide. Biodiesel is attracting attention as an eco-friendly fuel because it is more fuel efficient than petroleum and emits less exhaust gas. However, it is pointed out as a disadvantage that the production of biodiesel requires the use of food crops.

SAF was first developed in the 2000s and is evaluated as a more environmentally friendly fuel than biodiesel. SAF does not require the use of food crops because it uses renewable carbon sources such as waste cooking oil, waste paper, and wood pulp. In addition, SAF is evaluated as a more environmentally friendly fuel than biodiesel because it has better fuel efficiency and lower exhaust gas emissions than biodiesel.

The development status of biodiesel and SAF is as follows.

- *Biodiesel: In 2021, global biodiesel production is about 12 million tonnes. Of this, Europe produces about 4 million tons, the United States about 3 million tons, and Brazil about 2 million tons.*
- *SAF: As of 2021, world SAF production is about 1 million tonnes. Of these, the United States produces about 500,000 tons, Europe about 300,000 tons, and Singapore about 200,000 tons.*

Both biodiesel and SAF are attracting attention as eco-friendly fuels, but their high production cost is still pointed out as a disadvantage. However, as the production technology of biodiesel and SAF develops, production costs are expected to decrease, and the use of biodiesel and SAF is expected to expand in the future.

[Question 6] Please introduce the main approach for oxidation control of biodiesel and SAF.

[Google bard answer 6] The main approaches for oxidation control of biodiesel and SAF are as follows.

• *Addition of antioxidants*

Antioxidants prevent oxidation of biodiesel and SAF by inhibiting oxidation reactions. The most common antioxidant is tocopherol, a phenolic antioxidant. Tocopherol is added to biodiesel and SAF to inhibit oxidation and maintain fuel quality.

• *Removal of pro-oxidants*

Oxidation promoters are substances that promote oxidation reactions. Prooxidants are removed during the manufacturing process of biodiesel and SAF. Removal of pro-oxidants helps prevent oxidation of biodiesel and SAF and maintains fuel quality.

• *Improving storage conditions*

Biodiesel and SAF are prone to oxidation. Therefore, biodiesel and SAF need to prevent oxidation by adding antioxidants and improving storage conditions. Improving storage conditions helps prevent oxidation of biodiesel and SAF and maintains fuel quality.

• *Improving the manufacturing process of biodiesel and SAF*

The manufacturing process of biodiesel and SAF plays an important role in preventing oxidation of biodiesel and SAF. The manufacturing process of biodiesel and SAF should avoid oxidation of biodiesel and SAF by removing pro-oxidants and improving storage conditions.

Oxidation control of biodiesel and SAF plays an important role in maintaining the quality of biodiesel and SAF and improving the performance of the fuel. Various methods for controlling the oxidation of biodiesel and SAF are expected to promote the commercialization of biodiesel and SAF and accelerate the development of eco-friendly fuels.

[Question 7] Can the method of removing the dissolved oxygen inherent in biodiesel or SAF oxidation control be helpful?

[Google bard answer 7] Yes, in controlling the oxidation of biodiesel or SAF, the method of removing the dissolved oxygen inherent in it can be helpful.

Biodiesel and SAF are prone to oxidation. Oxidation degrades the quality of biodiesel and SAF, and causes degradation of fuel performance. Oxidation is caused by the oxygen contained in biodiesel and SAF. Therefore, in order to control the oxidation of biodiesel and SAF, it is important to remove the dissolved oxygen inherent in them.

The method to remove the inherent dissolved oxygen is as follows.

• *Distillation*
• *Absorption*

- *Ion exchange*
- *Separator*

> *These methods help control oxidation by removing dissolved oxygen inherent in biodiesel and SAF.*
>
> *For example, distillation is a method of boiling biodiesel and SAF to obtain an oxygen-free vapor. This method can effectively remove dissolved oxygen inherent in biodiesel and SAF.*
>
> *Adsorption is a method using materials that adsorb oxygen by reacting with biodiesel and SAF. This method can effectively remove dissolved oxygen inherent in biodiesel and SAF.*
>
> *Ion exchange is a method using a substance that reacts with biodiesel and SAF to convert oxygen into ions. This method can effectively remove dissolved oxygen inherent in biodiesel and SAF.*
>
> *Membrane is a method of using a membrane to separate biodiesel, SAF, and oxygen. This method can effectively remove dissolved oxygen inherent in biodiesel and SAF.*
>
> *Methods to remove inherent dissolved oxygen can help control the oxidation of biodiesel and SAF. These methods help maintain the quality of biodiesel and SAF and improve the performance of the fuel.*

[Question 8] Vacuum bubbling is a method of removing dissolved gases in a liquid by creating bubbles in a vacuum state. Could this method help control the oxidation of biodiesel or SAF? Do you have any examples of applying this method?

[Google bard answer 8] Yes, vacuum bubbling can help control the oxidation of biodiesel or SAF. Vacuum bubbling is a method of removing dissolved gases in a liquid by creating bubbles in a vacuum state. Biodiesel or SAF is prone to oxidation. Oxidation degrades the quality of biodiesel or SAF and causes the performance of the fuel to deteriorate. Oxidation is caused by oxygen contained in biodiesel or SAF. Therefore, it is important to remove the dissolved oxygen inherent in biodiesel or SAF to control oxidation. Vacuum bubbling can effectively remove dissolved oxygen inherent in biodiesel or SAF.

> *Examples of applying vacuum bubbling are as follows.*

- *The University of Alabama, USA, published a study on improving the oxidative stability of biodiesel using vacuum bubbling.*
- *The University of Rotterdam in the Netherlands has published a study on improving the oxidative stability of SAFs using vacuum bubbling.*

> *Vacuum bubbling is an effective method for controlling the oxidation of biodiesel or SAF. Vacuum bubbling helps to maintain the quality of biodiesel or SAF and improve the performance of the fuel.*

I would like to summarize the discussion on this issue with Google Bard's answers to my questions about the problem of oxidation control, which is an obstacle to the commercialization of biodiesel and SAF. The Korean experimental version of Google Bard is always comforting because it is a friend who always gives positive answers, but it should not be forgotten that the responsibility for verifying the answer lies with the questioner, as there is no place of responsibility for the answer. (Especially for the specific answer to my last question 8—definitely need a double-check.) Another answer regarding the applicability of vacuum bubbling says:

> Vacuum bubbling is an effective method for controlling the oxidation of biodiesel or SAF, but it also has disadvantages. Vacuum bubbling consumes a lot of energy and equipment is expensive. Also, dissolved oxygen removed through vacuum bubbling can be dissolved again in the liquid, so it must be used together with an oxidizing agent.

The method I suggest is low on energy costs, and the equipment does not have to be expensive at all. Therefore, there seems to be more and more interest in applying vacuum bubbling degassing to this field. It is expected that degassing using vacuum bubbling can be used to deaerate biofuels to solve technical bottlenecks for commercialization and contribute to supplying eco-friendly fuels.

3.3   ARTIFICIAL GILLS (POSSIBLE FUTURE HUMAN HABITAT: UNDERWATER DWELLING)

Artificial gills is an abstract word expression that reflects idealistic human desire to breathe underwater like a fish. Several names with even product names come out and disappear in this area, but interests are still there to make it happen. Though it is called "artificial gills," what I am proposing is the utilization of extracted gas, which has higher oxygen content of over 30%, out of vacuum bubbling process. Possible applications may cover leisure industry, especially with scuba diving equipment, submarine applications, and underwater habitat projects for the future.

According to Wikipedia [45], artificial gills are an unproven conceptualized device that allows humans to take oxygen from surrounding water, which is being introduced as a speculative technology that has not been demonstrated in a documented way. Almost all animals that naturally have gills are thermophilic (cold-blooded), requiring far less oxygen than thermostats (warm-blooded) of the same size [46]. This Wikipedia article concludes that, in practice, it is unclear whether humans can create useable artificial gills because of the large amounts of oxygen that must be extracted from water. In fact, a report [47] by the US Navy found that the average diver using a fully closed-circuit rebreather requires 1.5 L of oxygen per minute while swimming or 0.64 L per minute while resting. However, according to the educational reference site [48] of the Ministry of Education of the Republic of Korea, people generally breathe 15 to 16 times a minute, and about 0.5 L of air comes in and out at a time, so one can say that an average of 6.8 L per minute of air is involved. A simple calculation shows that the amount of oxygen in air is about 20.9%, so using 6.8 L of air per minute would give you about 1.42 L of oxygen per minute, which seems to fit well with the US Navy report. However, as a result of my actual measurement with a dissolved oxygen meter, the oxygen concentration of exhaled breath was about 18%, and in the case of long-suffering and deep exhalation, it was measured close to 16%. Therefore, in normal cases, 20.9% of air is inhaled, and about 3% of this oxygen is supplied to the body, and the rest is discharged out of the body. If so, the amount of net oxygen actually consumed through respiration can be estimated as follows. Breathing air volume × (intake air oxygen concentration – exhalation air oxygen concentration) = 6.8 lpm × (0.209 – 0.180) = 0.197 lpm. According to Wikipedia [49], when a person breathes, the body consumes oxygen and produces carbon dioxide. The basic metabolism requires about 0.25 L/min of oxygen at a breathing rate of about 6 L/min. A person can ventilate at a rate of 95 L/min but is known to metabolize only about 4 L/min of oxygen [50]. Metabolized oxygen is typically about 4% to 5% of the inspired volume at normal atmospheric pressure, or about 20% of the available oxygen in air at sea level. The air exhaled at sea level contains approximately 13.5% to 16% oxygen [51]. Here, it seems to be introduced close to the conservative value of my personal measurement experiment results. As such, the range of the amount of oxygen that a person needs for breathing is predicted to be quite wide, and the important point is whether this amount of oxygen can be easily extracted from water.

This is a big challenge, but the challenge of so-called "artificial gills," a device that allows humans to breathe underwater like fish, continues to this day. Most of the effort [52–54], however, is directed to extracting high concentrations of oxygen from water, which is far from actually allowing humans to directly breathe the extracted air. In addition, there have been reports of research on the possibility of using extracted dissolved oxygen for purposes other than breathing, namely, as a power source for underwater robots driven by fuel cells [55], or for applications to artificial lungs for medical use [56]. The meaning of "artificial gills" that I propose as an application field of vacuum bubbling is therefore a concept closer to a device that extracts high-oxygen-concentration air from underwater or an underwater habitat that applies it, rather than a personal portable underwater breathing device. Interestingly, recently, in Korea, there was a public announcement (announced on April 20, 2023) of the government-sponsored research support project the Disruptive Innovation R&D Design Planning Project, which included artificial gills as one of the subtopics. However, the content of the proposed topic is focused on the separation of dissolved oxygen in water, as mentioned earlier, and since there is no mention of specific requirements for the amount required for individual respiration, the current interest in this topic seems to be in the separation technology after all.

The oxygen separation technology demonstrated on the LikeAFiSH site [54] is also basically similar to the method I propose in that it is far from personally portable and is a system that supports underwater breathing, but the important difference is whether or not it is decompressed in advance.

Figure 3.10 is a high-oxygen-supply device using vacuum bubbling, which is a concept applied to an underwater shelter. The main components are vacuum bubbling vessel, vacuum equipment, service space, and carbon dioxide scrubber. First, the vacuum bubbling container and the carbon dioxide scrubber are filled with water, and the air stored in the high-pressure air tank is sent to the vacuum generator to make the container vacuum. Along with this, the pump-bubbler unit supplied by DC

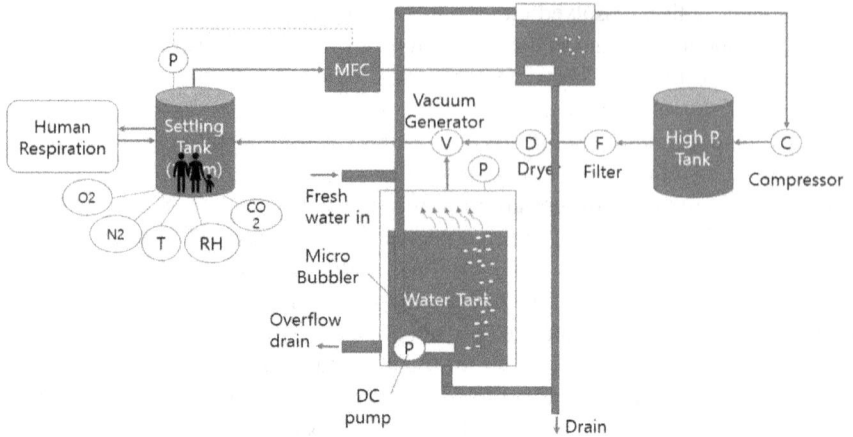

FIGURE 3.10 A conceptual view of an underwater respiration system using vacuum bubbling.

power creates bubbles and separates the dissolved gas, and the gas collected in the ullage at the top of the container is sent to the service space after being joined with the air flowing from the high-pressure tank in the vacuum generator. When air over the preset pressure is collected in the service space, the air in the service space is sent to the scrubber through the flow controller, and the air that has passed through the scrubber is introduced into the air compressor. The air inside the scrubber can be discharged intermittently as the volume of internal air is continuously increased by the air. The air stored in the high-pressure tank is controlled for dust, droplets, and humidity through filters and dryers.

In this system, air with a high oxygen concentration of about 30% or more generated in the vacuum bubbler is mixed with air with a low concentration and sent to the service area. In addition, the air in the service space passes through a scrubber with a bubbler, discharging carbon dioxide into water. The deaerated water and scrubber water are periodically or continuously returned to the water source through the outlet. The operation of these systems can be made continuously or intermittently as needed. This system can not only be applied as an underwater shelter for leisure sites but also will help improve indoor air quality by operating in conjunction with air conditioners in large buildings if sufficient water sources are available.

3.4 DESALINATION (MASS EVAPORATION WITH SOLAR RENEWABLES)

Desalination is an important engineering application to help people who need fresh water for their lives. Although desalination is already practiced on a large scale at many sites around the world, the expansion of its applications is limited by the high operating energy requirements. This is because current technology relies on high-energy-consumption processes, such as heating, vaporization, and condensation performed at high pressures. Current technologies can be classified into two groups based on their working principle: evaporation methods and membrane separations. Large-scale vacuum bubbling at moderately elevated temperatures has the potential to serve as an energy-saving alternative technology. Although it has to go through the existing evaporation and condensation processes in the same way, if it can be processed in vacuum pressure, the required operating temperature is not that high, and the required energy for heating and pumping can be supplied through renewable energy sources. Eventually, the desalination process can be carried out without relying on fossil energy any longer. This may be seen as a bit of a disadvantage compared to the heating method in terms of efficiency, but it also has a positive side in terms of more details. For example, in the conventional process of thermal desalination, condensation should be included after evaporation, but in the traditional high-temperature heating process, degassing of dissolved gases occurs naturally during the heating process, so there is no issue with the efficiency of the condensation process. However, in the vacuum evaporation process, if evaporation is performed without deaeration as a pretreatment, efficiency may become an issue in the condensation process due to low gas density and contribution of non-condensable gases. On the other hand, in the case of vacuum bubbling, a high level of degassing can be done in advance before full-scale vapor generation, so problems caused by the existence of non-condensable gas can be eliminated in advance. However, this book will be limited to introducing the applicable methodology because the desalination-related empirical tests have not yet been conducted. According to the desalination market forecast data (GWI 2016), the cumulative installed scale of desalination facilities around the world is expected to be 200 million m^3/day in 2030. Considering the electric energy consumption of 2.5–7 kWh/m^3 based on seawater RO data, the expected electric energy consumption for desalination is around 5 to 1.4 billion kWh/day. It can be seen that a considerable amount of energy is used in the desalination process. With the demand for desalination expected to continue to increase due to climate change, if vacuum bubble generation desalination can be put to practical use to suppress the use of fossil fuels in the desalination process, it can make a great contribution to carbon neutrality, I believe.

Before we discuss the topic of desalination, I would like to introduce two episodes. The first is an article published in *Scientific American* on July 23, 2008. The title is "Why don't we desalinate ocean water?" [57] and the answer is:

> The problem is that **the desalination of water requires a lot of energy**. Salt dissolves very easily in water, forming strong chemical bonds, and those bonds are difficult to break. Energy and the technology to desalinate water are both expensive, and this means that desalinating water can be quite expensive.

Second, when I searched "desalination conference 2023" just to monitor how active the research community on this subject is, I was able to realize the keen interest in this field. In one of them, Membrane Desalination 2023 (MEMDES2023), the following information was posted:

> Growing population, changing climate and increasing urbanisation require increasing freshwater supply and the sustainable water usage for industries and agriculture is of

top priority. One of the most viable solutions is to tap from the seawater and wastewater. Membrane technologies have been established as the golden standard for desalination to address the global deficit for clean water. However, it is still expensive, which is one of the main obstacles for prevalent adoption of membranes.

[58]

A really hot interest is being focused on desalination technology. This is because it is not only a matter of individual life but also of the fate of a community and a country, and the suffering of water shortage will spread exponentially if a sustainable solution is not presented.

3.4.1 WATER SCARCITY AND DESALINATION

The trend of the global drought index considering potential evaporation over the past 60 years announced by the National Research Council of Spain (Figure 3.11) [59] shows that many areas on the globe are already exposed to the threat of drought. In addition, Steven Solomon of the United States, who wrote *WATER: The epic struggle for wealth, power and civilization*, claims that

> The challenge of freshwater scarcity and ecosystem depletion is rapidly emerging as a defining factors of world politics and human civilization. A century of unprecedented freshwater abundance is eclipsed by a new age of acute disparities in water wealth, chronic insufficiencies, and deteriorating environmental sustainability across many of the most heavily populated parts of the planet. Just as oil conflicts played a central role in defining the history of the 1900s, so struggles for access to water resources is set to shape the destinies of societies and the new world order. Water is overtaking oil as the world's scarcest critical natural resource. But water is more than the new oil. Oil, in the end, is substitutable somehow by other fuel sources, or in extremis can be done without, but water's uses are pervasive, irreplaceable by any other substance, and utterly indispensable.

[60]

According to a report by the United Nations Environment Program [61],

> By 2025, due to climate change, more than half of the countries on the planet are expected to face a shortage of drinking water and other living water. Among the various methods to solve this problem, seawater desalination is the method that has received the most attention and has been put to practical use. Desalination technology by evaporation, which actually started in ships, developed rapidly in the 1950s and became applicable to large-scale seawater desalination plant commercial facilities. After the mid-1960s, membrane separation using reverse osmosis, a new desalination technology, started to be used.

These desalination technologies are classified as shown in Figure 3.11, according to the existence and state of phase change [62].

Since desalination is a topic of great public interest that goes beyond simple technology development and is connected to the future problems of mankind, interest in desalination is very hot, and a huge amount of research papers and reports is being

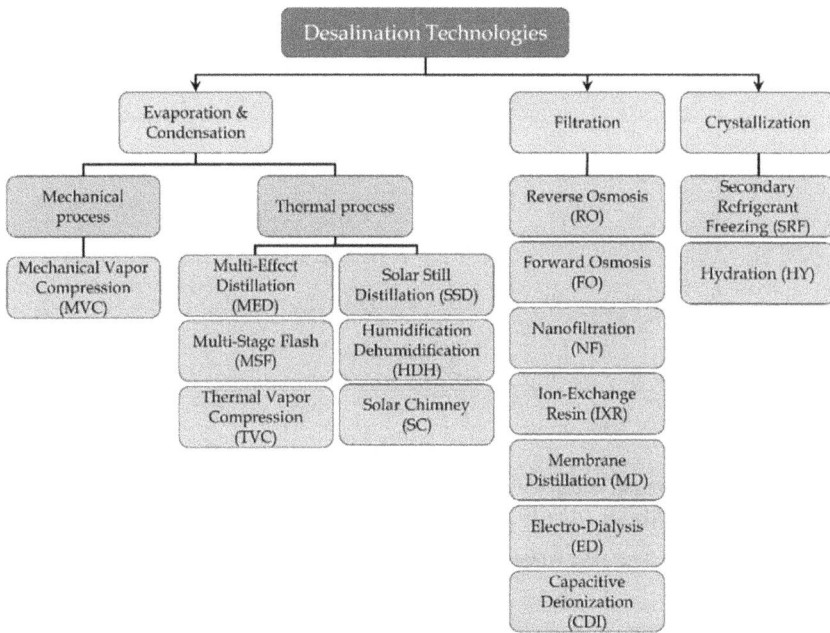

FIGURE 3.11   Classification of desalination technologies [62].

poured out. Approaching this topic has not been easy for an author without expertise in the field. However, the fact that desalination issues, including water shortage, are not the tasks of a group of experts in a specific business field or research field, but a universal problem of mankind, can only be of great interest to me as an outsider. I have consulted two accessible literatures to obtain a rough knowledge of the subject. One of them is "Desalination: a National Perspective" [63], published by the National Research Council in 2008, and another paper is an opinion paper from China introduced in the online media Frontiers in Energy Research in 2022 [64]. These two materials are interesting because they complement each other. The former has been published already 15 years ago, but it is the content of the technical report at the time when the supply of desalination had matured to some extent and membrane desalination facilities were in full swing in the United States. This report is interesting in that it has an introduction to the overall technology and an interim evaluation element, while the latter is a technology summary presented by Chinese researchers as one of the main consumers and suppliers of this technology. There will be many great materials on desalination technology, a big subject, but I express my gratitude by revealing that I tried to understand the current desalination technology through these two articles.

Though somewhat outdated, global desalination capacity as of 2006 has grown exponentially since 1960 to 42 million m³/day, as shown in Figure 3.12, with about 37 million m³/day of desalination capacity still operating as of 2006. Global desalination capacity almost doubled from 1995 to 2006 and continued to grow. At that time, nearly half (47%) of global online desalination capacity was in the Middle

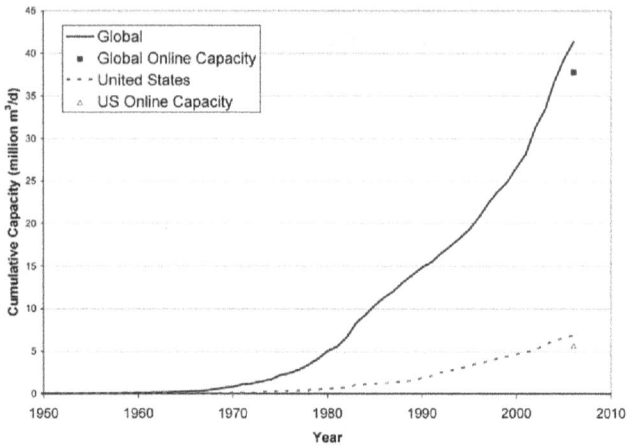

FIGURE 3.12 Cumulative capacity of installed desalination plants in the United States and worldwide from 1950 to 2006 [63].

Source: GWI. 2006b. 19th IDA Worldwide Desalting Plant Inventory. Oxford, UK: Media Analytics Ltd.

East, while North America, Europe, and Asia each held approximately 15% of global online desalination capacity (GWI, 2006b). Thermal and membrane processes are the two main processes in use worldwide. Major players include Siemens, GE, Veolia, Suez Degremont, Dow (Filmtec), Nitto Denko (Hydranautics), Toray, Pall, ITT, and Hyflux, which are estimated to have accounted for more than 75% of the global desalination membrane supply market. Recent updates on the cumulative global installed (online) capacity was 92.5 million m³/d in 2017 according to the posted information by IDA (International Desalination Association) [65]. Figure 3.13 shows similar but updated worldwide desalination capacity in terms of applied technology [66].

Water scarcity has spread from arid and semi-arid regions to wetter regions, and the availability of water to meet growing demand for household, agricultural, and environmental uses is becoming a growing problem. With recent severe droughts in many parts of the world and predictions of regional increases in the frequency and intensity of extreme events due to global climate change, desalination technologies have opened up the possibility of supplying large amounts of treated freshwater and providing a stable supply. Therefore, there is considerable interest in advancing desalination technology and concentrate management alternatives and hastening the time when their costs compete with those of other alternatives.

3.4.2 DESALINATION SYSTEMS IN ACTION

Five key elements of a desalination system for brackish or seawater desalination are (see Figure 3.14) [63]:

1. Intakes—the structure used to extract source water and convey it to the process system.

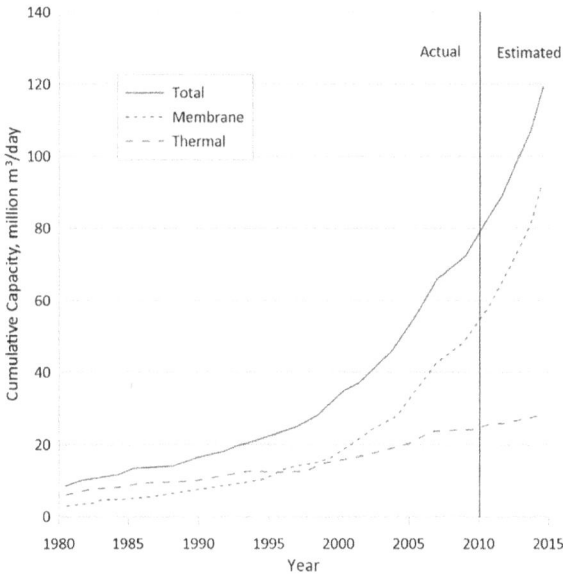

FIGURE 3.13 Cumulative installed worldwide desalination capacity in terms of applied technology [66].

Source: DesalData, Worldwide Desalination Inventory (MS Excel Format), 2013. Available from: DesalData. com (GWI/IDA) on June 2013.

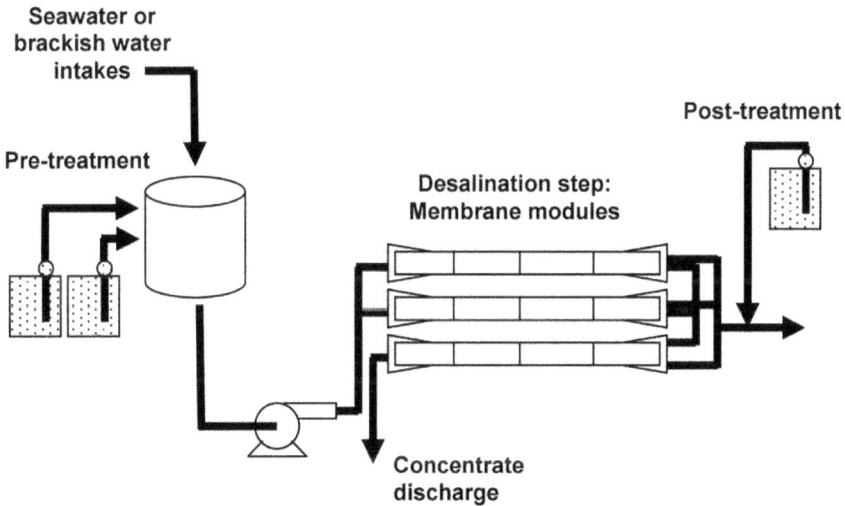

FIGURE 3.14 Key elements of a desalination system. Although shown here for a membrane-based system, these steps also describe the major components of non-membrane systems [63].

2. Pretreatment—removal of suspended solids and control of biological growth, to prepare the source water for further processing.
3. Desalination—the process that removes dissolved solids, primarily salts and other inorganic constituents, from a water source.
4. Post-treatment—the addition of chemicals to the product water to prevent corrosion of downstream infrastructure piping.
5. Concentrate management—the handling and disposal or reuse of waste residuals from the desalination system.

Pretreatment is generally required for all desalination processes to ensure that raw water components do not degrade the performance of the desalination plant. The desalination process refers to the step in which dissolved solutes are substantially removed from the feedwater to produce the desired product water. Existing technologies include the more commonly used membrane, thermal, and ion exchange processes. Membrane processes, including reverse osmosis (RO), nanofiltration (NF), electrodialysis (ED), and reverse electrodialysis (EDR), account for 56% of global desalination online capacity. The basic concept of thermal distillation is to generate water vapor by heating saline water. When this vapor is directed to a cold surface, it can condense into liquid water that originally contains little salt. Water boils at 100°C under atmospheric pressure, but thermal processes can also be designed to boil water in a series of vessels operated at successively lower temperatures and pressures. Thermal processes such as multi-stage flash (MSF) and multiple effect distillation (MED) account for 43% of global desalination online capacity.

So what are the energy costs for these technologies? Most public data do not clearly disclose data on initial installation costs, data on operation and maintenance, and costs on pre- and post-treatment. For fair judgment, the full cycle cost estimation of the process including these data should be applied to this field as well. However, for my own and the readers' reference, some excerpts from the published data in the references are introduced in Tables 3.2 and 3.3. It is hoped that this will help in recognizing the approximate energy consumption, but since there is no presentation of details such as the calculation method, excessive expectations about the accuracy or representativeness of the data are not recommended.

Although RO technology appears to be maturing, several major challenges remain, including membrane fouling leading to increased energy use and lower resistance to chlorine and other oxidants. . .. Maximum recovery is limited by the mechanical pressure limitations of the material on the membrane element. . .. Fewer and larger membrane elements can reduce operating and maintenance requirements while reducing overall capital costs through economies of scale and the need for fewer components (e.g., piping, connections). Nanofiltration – Similar to the RO process, the NF process also uses a semi-permeable membrane and a hydraulic driving force ranging from approximately 50-250 psi. Three major thermal processes have been commercialized: MSF distillation, MED and MVC, and each is a mature and powerful technology. MSF and MED processes require both thermal energy (usually steam) and electrical energy.

The government report quoted, although not up-to-date, helps us understand the current technology and challenges of desalination. So what about the relatively

TABLE 3.2

Comparison of Predominant Seawater Desalination Processes

	Seawater RO	MSF	MED (with TVC)	MVC
Operating temperature (°C)	<45	<120	<70	<70
Pretreatment requirement	High	Low	Low	Very low
Main energy form	Mechanical (electrical) energy	Steam (heat)	Steam (heat and pressure)	Mechanical (electrical) energy
Heat consumption (kJ/kg)	NA	250–330	145–390	NA
Electrical energy (kWh/m³)	2.5–7	3–5	1.5–2.5	8–15
Product water quality (TDS mg/L)	200–500	<10	<10	<10

Source: Quoted from [63] with permission.

TABLE 3.3

Comparison of Predominant Brackish Water Desalination Processes

	Brackish Water RO	ED/EDR	NF
Operating temperature (°C)	<45	<43	<45
Pretreatment requirement	High	Medium	High
Electrical energy (kWh/m³)	0.5–3	~0.5 per 1,000 mg/L of ionic species removed	<1

Source: Quoted from [63] with permission.

recent discussion of desalination? Let us take a look at the latest technology description [64].

Wang and Huo [64] classified desalination technologies into thermal drive, mechanical drive, and electric drive, depending on the driving energy required during the desalination process. It introduces MED (multiple effect distillation) and MSF (multi-stage flash) as thermally driven desalination technologies and mentions the low energy consumption and flexible installation capacity of reverse osmosis (RO) during membrane process as mechanically driven desalination technology. Electrically driven desalination technology, in which ions pass through a selective exchange membrane under a direct current electric field to obtain fresh water, consumes a lot of direct electrical energy and reports only 2% of global desalination capacity. In addition, new desalination technologies such as membrane distillation (MD), which combines thermal energy with membrane separation, are seen as promising as they can be powered by low-grade heat rather than electricity.

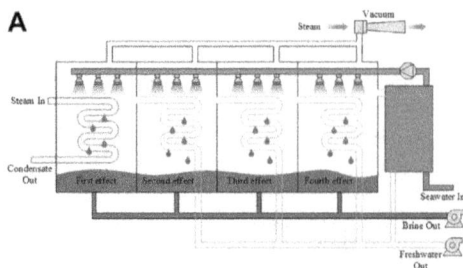

FIGURE 3.15 Schematic of the seawater desalination process: (A) MED (multiple effect distillation) method [64].

Now, let us look at how each works. The operating principle of multiple effect distillation (MED), Figure 3.15, is understood as follows. The pretreated raw water passes through the last condenser in the process, and after being heated, it is sprayed at high pressure in vacuum containers connected in a plurality of series. This can be seen to be a similar situation as the liquid droplet sprayed in the spray-type deaerator is heated and evaporated under a higher temperature. The vapor evaporated from the raw water is transmitted to the inside of the next-stage vacuum vessel under a lower pressure connected to the pipe and condenses while heating and vaporizing the droplets sprayed from the corresponding stage. At this time, if the pressure is kept low as the number of treatment stages increases, the saturated steam temperature of the water will be lowered so that the vaporization can occur smoothly at a lower temperature. In this way, condensation proceeds while maintaining a lower temperature and pressure in turn from the container (no. 1), where raw water evaporates first by steam supplied from an external boiler. This system consists of a steam supply line supplied by an external boiler, a pressure vessel constituting a plurality of stages, a condenser coil installed inside each vessel, a high-pressure spray nozzle installed upstream of the raw water side, and a vacuum facility for decompression of the vessel. The illustration shows the use of a steam ejector that uses steam to create a vacuum condition.

In the late 1970s, IDE developed low-temperature multiple-effect distillation (LT-MED) technology using low-grade steam at 50–70°C to reduce desalination costs and alleviate corrosion and scaling problems, resulting in freshwater production costs of $0.738/ton (Liu et al., 2021).

The multi-stage flash (MSF) method, Figure 3.16, is based on the principle of flash evaporation. Seawater is heated by steam and enters the flash evaporator, where the hot seawater is rapidly vaporized because the pressure in the flash evaporator is lower than the saturated vapor pressure. The resulting steam is condensed into fresh water while heating the forward seawater, and the remaining seawater is sent to the next flash evaporator, where it is vaporized at a lower pressure. MSF technology is reliable and scalable but has a high heat demand, so it is often combined with thermal power plants for cogeneration of fresh water and electricity. And the cost per ton of water produced in this combined system is about $1.023 (Semiat, 2008).

Wang and Huo [64] expressed expectations for renewable energy–based desalination, especially desalination using vacuum, as an effective way to solve problems

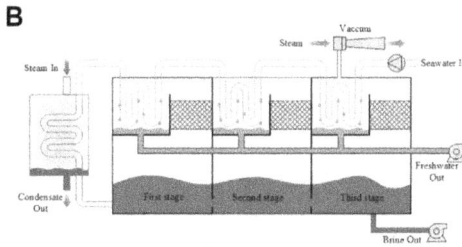

FIGURE 3.16 Schematic of the seawater desalination process: (B) MSF (multi-stage flash) method [64].

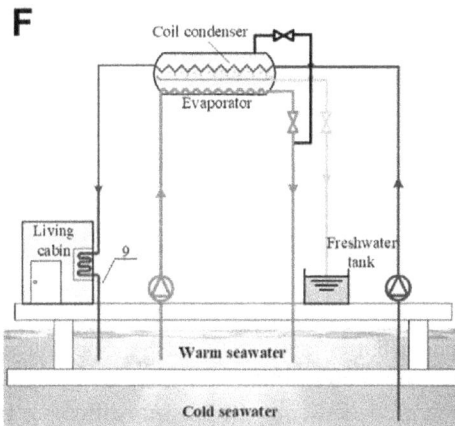

FIGURE 3.17 Schematic of the seawater desalination process: (F) low-pressure flash evaporation system powered by ocean thermal energy [64].

such as high investment cost, high energy consumption, and environmental pollution of traditional seawater desalination technology. It is instantly evaporated at low temperature by a vacuum pump, and the steam produced is exchanged with cold deep seawater and condensed into fresh water. However, it is pointed out that the non-condensable gas (NCG) dissolved in the seawater is gradually released during the flashing process, and the condensation process is then hindered by the NCG, which reduces the freshwater yield. Natural vacuum technology is based on the Torricelli phenomenon, in which warm seawater at 30°C evaporates from a water column at a height of 10.33 m under natural conditions (Figure 3.17). The power consumption and recovery rate achieved through this technology were reported to be 0.126 kWh/kg and 1.5%, respectively. The siphon flash evaporation desalination system (SFEDS) using ocean thermal energy, which is close to MSF concept-wise but relying on ocean thermal energy rather than on steam, is introduced by Jin et al. [67] and Wang et al. [68]. Figure 3.18 is a schematic diagram of the SFEDS process. This system consists of four parts: surface seawater circulation, deep seawater circulation, vacuum system, and storage system. The vacuum system mainly consists

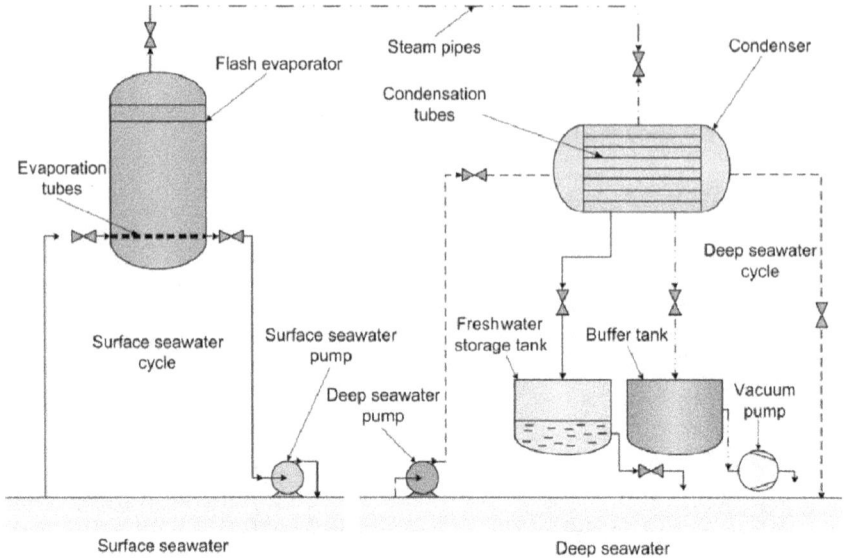

Fig. 1. Flow chart of SFEDS.

FIGURE 3.18 System diagrm of siphon flash evaporation desalination system [67].

of a vacuum pump and a buffer tank, and a flash evaporator is placed at a certain height, where natural evaporation of surface seawater takes place. The generated steam is absorbed into the condenser by the pressure difference between the evaporator and the condenser and condensed into fresh water by the cold deep water. Due to a certain vacuum difference between the condenser and the evaporator, the vapor is continuously absorbed into the condenser. The interesting part of this model is that it relies on natural evaporation by reduced pressure without a separate external heating source and achieves condensation through cold deep water.

Then, what about the evaluation of desalination technology? A report from 2008 [63] noted:

> While the major desalination technologies currently in use are generally efficient and reliable, cost and energy requirements remain high. Ongoing research efforts are driven by the need to reduce costs or overcome operational limitations of the process, such as reducing membrane fouling or increasing energy efficiency. Existing desalination technologies will continue to show incremental improvements, but current technologies are relatively mature and the practical limit for additional energy savings through advances in RO membranes is around 15%, so existing desalination processes can be improved, replaced or mainstreamed. Alternatives to the main desalination technologies continue to be explored to fill niche applications where the technology is unavailable.

And:

> [N]o desalination process can overcome the thermodynamic limit of desalination in terms of energy use. Nonetheless, efforts to approach the thermodynamic energy limit

more closely or to find ways to power the desalination process with cheap energy sources such as low-grade heat. Research into alternative desalination technologies is ongoing.

Here are some of Wang and Huo's conclusions in 2022, decades later: "The past traditional seawater desalination technology has problems such as high investment cost, high energy consumption, and environmental pollution, and renewable energy-based seawater desalination technology is an effective way to solve these problems." It has been declared, but it is reported that the proven results are unfortunately not yet up to expectations. One of the recent review papers for an overview of desalination technology is the work of Curto et al. [62]. They provide one of the most recent and comprehensive review of the technology in this field by covering as much as possible all known technologies in practice. One of their important conclusions says that RO is the best available technology so far and is easily connected to renewable energy. Interestingly, however, they did not consider the approaches using vacuum technologies serious. One piece of literature I would like to add before ending this story is the paper by Mansour and Muller [69] on flash evaporation. Flash evaporation seems promising when one is considering fast-phase separation, and it has been used for conventional desalination processes. The application of flash evaporation may be extended to lower pressure situations like VBD and SFEDS, and it may serve a critical role for it. It seems to be a conceivable task to deaerate water using vacuum bubbling, which will be a prerequisite for high condenser performance, followed by flash evaporation desalination, even though the way one makes vacuum condition may come out in different fashion.

3.4.3 Vacuum Bubbling Desalination with Solar Renewables

Vacuum bubbling desalination (VBD) process, shown in Figure 2.46 in Chapter 2, is close to MSF process in that water at its near phase-change condition is expelled to a lower-pressure region to make the vapor separated from liquid and then to be condensed. Figure 3.19 illustrates the conceptual model for possible application to desalination process. Then, what are the differences?

The biggest issue is in that it does not rely on steam for heating of saline water to be desalinated. In typical MSF process, the use of steam from the boiler system is a must, but in a VBD system, moderate heating from solar water is good enough. Instead, we rely on much lower pressure for phase change. Let us take a look at the working condition, selected for presentation, in Figure 3.19. We assume the water temperature of saline water to be $T_{w,1} = 40°C$ and that in the condenser to be $T_{w,5} = 20°C$, respectively. In this situation, the pressure in the vessel $p_1 \le p_{sat} = 7.4\,kPa$ needs to be secured in order for the generated vapor bubbles to neither immediately shrink nor collapse back to water before it is supposedly condensed. The nozzle downstream pressure, p_4, will probably be higher than p_1 when the nozzle is submerged down from the water surface by depth Δh, as illustrated in Figures 3.19(a) and (b). The working principle of VBD can be described as follows:

1. Once warm water is filled up, the ullage air is vented out down to the desired level, to say, 5 kPa. This is lower than p_{sat}, and the preferred state of water is in the form of vapor.

2. Once the submersible pump is on, they develop through the venturi nozzle and generate bubbles. Supposedly, the pressure at the nozzle throat, p_3, is much lower than p_{sat} so that massive vapor bubble may be generated. The survival of these vapor bubbles is believed to depend on the nozzle downstream pressure, p_4. Therefore, the desired pressure level is such that $p_4 \leq p_{sat}$. Once bubbles survive at the nozzle downstream, the condition thereafter is favorable because of the uprising movement of bubbles to where the pressure is getting lower till they eventually arrive at the water surface condition of p_1.

3. The vapors are then transported to the condenser section, where temperature is lower, $T_5 = 20°C$. Vapors are supposedly condensed to water on the condenser surface and drop down to the container. This process is briefly illustrated in Figure 3.19(b).

This is a basic concept of the VBD process; however, details may be of importance.

[Question 1] How much evaporation is going to happen per given kWh?

[My answer 1] Assuming that we already reached the phase-change state, in terms of pressure under a given temperature condition ($T_{w,1} = 40°C$), $p \leq p_{sat}$, and assuming the energy is solely used to increase the enthalpy of the concerned mass, the expected amount of phase change of water to be 2,406 kJ/kg = 0.668 kWh/kg, which is believed to be the thermodynamic limit and needs to be used as a reference value. So the answer is that 1.5 kg of water or 1.5 L of water per kWh input is ideally expected. In this regards, the working performance of flash evaporators need to be referred.

[Question 2] What are the possible issues with higher efficiency?

[My answer 2] On top of the vapor generation, condensing is also important. The gas may include non-condensing-gases (NCG), too, which may degrade the condenser performance, plus the density is so low the condensation efficiency should be of concern. Depending on how the vacuum system is arranged, the final performance and the expense will be determined. The first issue of degassing NCG can be treated almost perfect, but it will charge additional energy cost. The implementation of vacuum system at the downstream of condenser space may minimize the power needed for the process operation. Of course, how good it works depends on the condenser design and performance.

[Question 3] Are there possibility of pre-condensing before vapor arrives at condenser section?

[My answer 3] Yes. The state of inner space is barely O.K. for maintaining vapor state, but if the outside temperature is lower than $T_{w,1} = 40°C$, vapor close to the walls that is exposed to outside condition may start condensing in the evaporation vessel. This should be considered carefully.

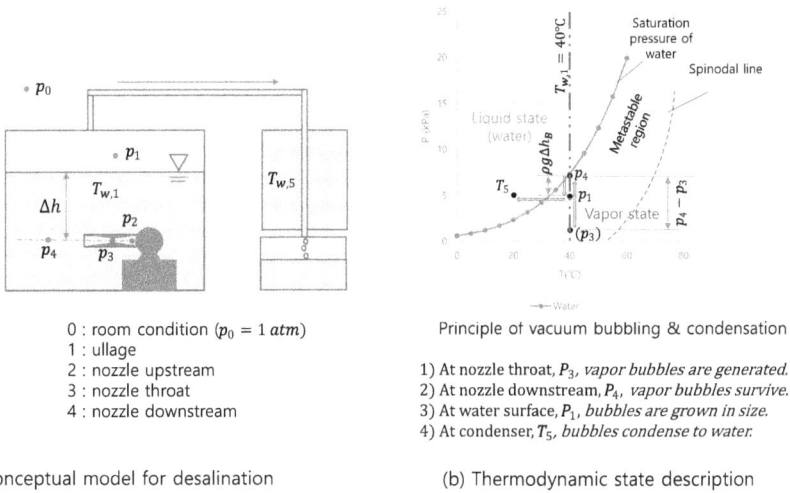

0 : room condition ($p_0 = 1\ atm$)
1 : ullage
2 : nozzle upstream
3 : nozzle throat
4 : nozzle downstream

Principle of vacuum bubbling & condensation

1) At nozzle throat, P_3, vapor bubbles are generated.
2) At nozzle downstream, P_4, vapor bubbles survive.
3) At water surface, P_1, bubbles are grown in size.
4) At condenser, T_5, bubbles condense to water.

(a) Conceptual model for desalination

(b) Thermodynamic state description

FIGURE 3.19   A conceptual view of a desalination system based on massive vapor bubbling along with solar renewables.

[Question 4] How come you are dealing with so tiny amount of input power rather than using more powerful pump? Would it not be more beneficial?

[My answer 4] A good question. I believe more powerful agitators may be used to make bubbles for degassing purposes. However, when you deal with high vacuum, the sealing is a critical factor. Also the beauty of dealing with lower pressure with water, lies in that the needed power is not much either, compared to the case of generating cavitation bubbles on the back side of ship propellers. Sealing that can support continuous operation needs to be guaranteed before one attempts VBD.

REFERENCES

[1] "Deaerator," Wikipedia – The Free Encyclopedia. https://en.wikipedia.org/wiki/Deaerator
[2] Google Patent US1914166A, Apparatus for treating liquids. Retrieved August 21, 2023.
[3] Stickle History, "History—stickle steam specialties." https://sticklesteam.com/history/. Retrieved January 14, 2024.
[4] Spirax Sarco, "The feedtank and feedwater conditioning." https://www.spiraxsarco.com/learn-about-steam/the-boiler-house/the-feedtank-and-feedwater-conditioning#article-top. Retrieved January 14, 2024.
[5] OSTI Abstract, "Vacuum deaeration in waterflood operations (Conference)." Osti.gov. Retrieved August 21, 2023.
[6] "Deaerator plant (image)." https://en.wikipedia.org/wiki/Deaerator#/media/File:Open_deaerator_plant.jpg. Retrieved August 21, 2023.
[7] "Tray-type deaerator (image)." https://upload.wikimedia.org/wikipedia/commons/2/23/Deaerator-en.svg. Retrieved August 21, 2023.

[8] "Spray-type deaerator (image)." https://en.wikipedia.org/wiki/Deaerator#/media/File:SprayDeaerator-en.svg. Retrieved August 21, 2023.

[9] "Rotating disc deaerator (image)." https://en.wikipedia.org/wiki/Deaerator#/media/File:Vacuum_rotating_disc_deaerator.jpg. Retrieved August 21, 2023.

[10] Membrane Deaeration, "Cannon Artes homepage." www.cannonartes.com/systems-equipment/deaeration/membrane-deaeration/. Retrieved August 21, 2023.

[11] P. Peterson, Membrane Technology: The Future of Deaeration of Injection Water. The Produced Water Society Seminar, 2016. https://producedwatersociety.com/wp-content/uploads/2021/07/00185_3M-Presentation.pdf. Retrieved August 21, 2023.

[12] "A high level of consistency and control, 3M™ Liqui-Cel™ membrane contactors." https://multimedia.3m.com/mws/media/2091084O/3m-spsd-liqui-cel-general-brochure.pdf. Retrieved August 21, 2023.

[13] Ministry of Foreign Affairs of the Republic of Korea, (Joint Press Release) "Intergovernmental panel on climate change, approval of the 6th assessment report," March 20, 2023. www.mofa.go.kr/www/brd/m_4080/view.do?seq=373483

[14] IRENA, World Energy Transitions Outlook 2022: 1.5°C Pathway, Executive Summary of IRENA. International Renewable Energy Agency, Abu Dhabi, 2022. www.irena.org/publications

[15] Eurowater, "Vacuum deaerator – remove oxygen and carbon dioxide from make-up water." eurowater.com. Retrieved August 21, 2023.

[16] "Vacuum deaerator." www.lm-tech.kr/Vacuum_Deaerator. Retrieved August 21, 2023

[17] "Vacuum deaeration, EWT Water Technology." ewt-wasser.de. Retrieved August 21, 2023.

[18] "Whittier™ vacuum deaerators – industrial grade degasification." www.veoliawater-technologies.com/en/technologies/whittier-vacuum-deaerators. Retrieved August 21, 2023.

[19] American Water Chemicals, "Deaeration systems." www.membranechemicals.com/water-treatment/deaeration-systems/. Retrieved August 21, 2023.

[20] "Zero gas vacuum deaerator, Cannon Artes." https://www.cannonartes.com/systems-equipment/deaeration/zerogas-vacuum-deaerators/. Retrieved January 14, 2024.

[21] "Vacuum deaerator (DEB)." https://bachiller.com/en/vacuum-deaerator/. Retrieved August 21, 2023.

[22] "PerMix vacuum deaerator," PerMix. www.permixmixers.com/liquid-mixers/permix-vacuum-deaerator/. Retrieved August 21, 2023.

[23] "Beverage deaeration," TechniBlend. www.techniblend.com/products/beverage-deaeration/. Retrieved August 21, 2023.

[24] "High vacuum water deaeration V2WD," corosys. https://bbt.corosys.com/en/solutions/degassing/high-vacuum-water-deaeration-v2wd/. Retrieved January 14, 2024.

[25] "Deaeration," Enhydra Ltd. https://enhydra.co.uk/en-wp/wp-content/uploads/2015/11/Deaeration_2015_v3.pdf. Retrieved January 14, 2024.

[26] "Homepage of Eurowater." www.eurowater.com/en/applications

[27] Corosys Beverage Technology. https://bbt.corosys.com/en/solutions/degassing/high-vacuum-water-deaeration-v2wd/

[28] L. Kennedy, "What happened to TWA flight 800? July 12, 2021. www.history.com/news/twa-flight-800-crash-investigation. Retrieved July 29, 2023.

[29] C. Yan, B. Xueqin, L. Guiping, S. Bing, Z. Yu, L. Zixuan, "Experimental study of an aircraft fuel tank inerting system," Chinese Journal of Aeronautics, 2015, 28(2), pp. 394–402.

[30] S. M. Summer, Mass Loading Effects on Fuel Vapor Concentration in an Aircraft Fuel Tank Ullage, Report No. DOT/FAA/AR-TN99/65. Federal Aviation Administration, Atlantic, NJ, 1999.

[31] S. M. Summer, "Limiting oxygen concentration required to inert jet fuel vapors existing at reduced fuel tank pressures-final phase," Report No. DOT/FAA/AR-04/8, 2004.

[32] R. Langton, C. Clark, M. Hewitt, L. Richards, Aircraft Fuel Systems. John Wiley & Sons, Chichester, UK, 2009, pp. 225–234.

[33] D. E. Smith, "Fuel tank inerting systems for civil aircraft," M.S. Thesis, Colorado State University, 2014.

[34] P. F. Dunn, F. O. Thomas, J. B. Leighton, D. Lv, "Determination of Henry's law constant and the diffusion and polytropic coefficients of air in aviation fuel," Fuel, 2011, 90, 1257–1263.

[35] ASTM D2779-92, Standard Test Method for Estimation of Solubility of Gases in Petroleum Liquids (Reapproved), 2012.

[36] W. M. Cavage, "The effect of fuel on an inert ullage in a commercial transport airplane fuel tank," Final Report No. DOT/FAA/AR-05/25, 2005.

[37] I. Martinez, "Solubility data for aqueous solutions." http://imartinez.etsiae.upm. es/~isidoro/dat1/Solubility%20data.pdf. Retrieved July 30, 2023.

[38] R. Sander, "Compilation of Henry's law constants (version 4.0) for water as solvent," Atmospheric Chemistry and Physics, 2015, 15, pp. 4399–4981.

[39] D. Tromans, "Temperature and pressure dependent solubility of oxygen in water: A thermodynamic analysis," Hydrometallurgy, 1988, 48, pp. 327–342.

[40] Biodiesel. https://namu.wiki/w/%EB%B0%94%EC%9D%B4%EC%98%A4%20%EB% 94%94%EC%A0%A4

[41] N. Kumar, "Oxidative stability of biodiesel: Causes, effects and prevention," Fuel, 2017, 190, pp. 328–350.

[42] L. Longanesi, A. P. Pereira, N. Johnston, C. J. Chuck, "Review: Oxidative stability of biodiesel: Recent insights," Biofuels, Bioproducts and Biorefining, 2022, 16, pp. 265–289. http://doi.org/10.1002/bbb.2306

[43] M.-E. Lee, I.-H. Hwang, K.-K. Kim, B.-K. Na, "Review on the oxidation stability of biodiesel," Journal of Oil & Applied Science, 2018, 35(4), pp. 1013–1030 (in Korean).

[44] K.-I. Min, E.-S. Yim, C.-S. Jung, J.-K. Kim, B.-K. Na, "Study of oxidation degradation of automotive diesel on storage circumstances," Proceedings of 2012 KSAE Annual Conference, pp. 372–379, KSAE12-B0076.

[45] "Artificial gills," Wikipedia – The Free Encyclopedia. https://en.wikipedia.org/wiki/ Artificial_gills_(human)

[46] Howstuffworks express. https://web.archive.org/web/20071111030608/http://express. howstuffworks.com/mb-gills.htm

[47] M. E. Knafelc, Oxygen Consumption Rate of Operational Underwater Swimmers. ADA205331. Navy Experimental Diving Unit, Panama City, FL, 1989. https://web. archive.org/web/20081122050431/http://archive.rubicon-foundation.org/7406

[48] "What is breathing?" (in Korean). Ministry of Education, Republic of Korea. https://m. blog.naver.com/moeblog/220437084981

[49] "Rebreather," Wikipedia – The Free Encyclopedia. https://en.wikipedia.org/wiki/ Rebreather

[50] J. T. Joiner (ed.), NOAA Diving Manual, Diving for Science and Technology, 4th ed. National Oceanic and Atmospheric Administration, Office of Oceanic and Atmospheric Research, National Undersea Research Program, NOAA Diving Program (U.S.), Silver Spring, MD, 2001 (CD-ROM prepared and distributed by the National Technical Information Service (NTIS) in partnership with NOAA and Best Publishing Company).

[51] P. S. Dhami, G. Chopra, H. N. Shrivastava, A Textbook of Biology. Pradeep Publications, Jalandhar, 2015, p. V/101.

[52] J. Lee, P. W. Heo, T. Kim, "Theoretical model and experimental validation for underwater oxygen extraction for realizing artificial gills," Sensors and Actuators A: Physical, 2018, 284, pp. 103–111.

[53] P. W. Heo, "Artificial gill technology," Machine and Material, 2010, 22(10), pp. 110–119.

[54] LiKEAFiSH Home Page, "Technology—like a fish." https://www.likeafish.biz/. Retrieved January 14, 2024.

[55] I. Ieropoulos, C. Melhuish, J. Greenman, "Artificial gills for robots: MFC behavior in water," Bioinspiration & Biomimetics, 2007, 2, pp. S83–S93.

[56] G. B. Kim, S. J. Kim, M. H. Kim, C. U. Hong, H. S. Kang, "Development of a hollow fiber membrane module for using implantable artificial lung," Journal of Membrane Science, 2009, 326, pp. 130–136.

[57] P. Gleick, "Why-don't-we-get-our-drinking-water-from-the-ocean?" www.scientificamerican.com/article/why-dont-we-get-our-drinking-water-from-the-ocean/

[58] www.elsevier.com/events/conferences/desalination-using-membrane

[59] S. Begueria, B. Latorre, F. Reig, S. M. Vicente-Serrano, "Global drought monitor." https://spei.csic.es/map/maps.html#months=1#month=5#year=2023

[60] S. Solomon, Water: The Epic Struggle for Wealth, Power and Civilization. HarperCollins e-books, 2010.

[61] UNESCO World Water Assessment Programme, "Water for people, water for life: The United Nations world water development report; a joint report by the twenty-three UN agencies concerned with freshwater (kor)," UNESCO, 2003. https://unesdoc.unesco.org/ark:/48223/pf0000129726_kor

[62] D. Curto, V. Franzitta, A. Guercio, "Review: A review of the water desalination technologies," Applied Sciences, 2021, 11, p. 670. https://doi.org/10.3390/app11020670

[63] National Research Council, Desalination: A National Perspective. The National Academies Press, Washington, DC, 2008. https://doi.org/10.17226/12184.

[64] J. Wang, E. Huo, "Opportunities and challenges of seawater desalination technology," Frontiers in Energy Research, Opinion, 2022, 10, article 960537. www.frontiersin.org, http://doi.org/10.3389/fenrg.2022.960537

[65] International Desalination Association, "Desalination at a glance," 2011. https://idadesal.org/wp-content/uploads/2021/06/desalination-at-a-glance.pdf

[66] L. O. Villacorte, S. A. A. Tabatabai, N. Dhakal, G. Amy, J. C. Schippers, M. D. Kennedy, "Algal blooms: An emerging threat to seawater reverse osmosis desalination," Desalination and Water Treatment, 2015. http://doi.org/10.1080/19443994.2014.940649/ (original source from DesalData, Worldwide Desalination Inventory (MS Excel Format), 2013. DesalData.com (GWI/IDA)).

[67] Z. J. Jin, H. Ye, H. Wang, H. Li, J. Y. Qian, "Thermodynamic analysis of siphon flash evaporation desalination system using ocean thermal energy," Energy Conversion and Management, 2017, 136, pp. 66–77.

[68] L. Wang, X. Ma, H. Kong, R. Jin, H. Zheng, "Investigation of a low-pressure flash evaporation desalination system powered by ocean thermal energy," Applied Thermal Engineering, 2022, 212, p. 118523.

[69] A. Mansour, N. Muller, "A review of flash evaporation phenomena and resulting shock waves," Experimental Thermal and Fluid Science, 2019, 107, pp. 146–168.

4 Concluding Remarks

Vacuum bubbling can be seen as one of the fields that have not received much attention in the past and have not been studied in depth. Much of vacuum bubbling is actually not considered a new technology; there is no great technical challenge, but the expectation for the effect is not high, so it is a field that has not been considered as a field that can create high added value in industry. So far, most researchers have been interested in the results of solubility behavior governed by Henry's law, but the reason this field has come to me in particular is from the curiosity about what happens after the solubility limit. This is because when we tried to lower the limit of solubility by lowering the pressure, we faced a limiting situation, which is a phase change. When the solution undergoes a phase change, solubility itself, which is defined as the amount of solute that can be dissolved in a liquid solution, is no longer defined, so many people seem not to have paid attention to the situation beyond that. However, in the background of the existing technology, it is not that there was no interest in the situation after the phase change. For example, great achievements have already been made regarding the behavior of steam at high temperatures. And the results are actually applied. In the thermal degassing method, degassing by steam is finally carried out. The expected level of outgassing is known to be 5 ppb, which is probably the highest publicly required level of outgassing. (There are no published data on the degassing method and performance used in some semiconductor processes.) It was already known that the concentration of steam becomes very low due to the effect of volume expansion when vaporized through phase change, whether at low pressure or high pressure. That is why degassing using a steam sparger is carried out without exception in the field, but the need for vapor bubbles generated at room temperature and low pressure may not have been thought as such, at least until now. I emphasize that the vacuum bubbling in this work means bubbling, which is a separation phenomenon of supersaturated solutes within the range of application of Henry's solubility law, as the first step, and the additional contribution by vapor bubbles generated below the saturated vapor pressure as the second step.

What makes vacuum bubbling attractive is that the energy consumption involved is significantly less than heating. In general, heating and depressurization can be thought of as two distinct approaches that can be selected as paths to reach the phase-change state. These two paths target changes for the two independent state variables, temperature and pressure. An interesting point is that decompression, unlike heating, does not directly apply energy to the system but changes the surrounding conditions. In order to heat water, the energy level of the water itself must be increased through heating, but depressurization, on the other hand, has the effect of changing the surrounding environment around water rather than making any changes to the water itself. Although the energy required for heating is not relatively large compared to the latent heat of vaporization, the absolute amount of energy is still huge. In contrast,

DOI: 10.1201/9781003374626-4

the amount of work required for depressurization, though it depends on the initial volume, is estimated to be orders of magnitude less than heating energy to reach the phase-change state. In particular, this difference contrasts dramatically with thermal degassing, which requires both heating and evaporation of the entire water to be treated. In particular, it is to be noted that two-step depressurization seems the key to minimize the energy required for bubble generation.

In this work, vacuum bubbling is presented in two clearly distinct phases, one governed by solubility and one governed by phase change. The reason it is called a phase is that it is usually dominated by solubility first as the pressure decreases from atmospheric pressure, and then vapor bubbles will be created by phase change under lower pressure. In particular, discussions related to the generation and behavior of room-temperature vapor bubbles include topics that have not been studied much in the traditional academic community. In this work, an analysis of the vapor bubble generation method and conditions was attempted, and the experimental results were presented. Above all, the most important expected result is related to the degassing performance. For the oxygen concentration in vapor bubbles under vacuum is expected to be lower than that of hot steam, it will acquire a very unique position in the field of degassing. A result of the degassing test in water through vapor bubbles has already demonstrated that the concentration of dissolved oxygen could reach the minimum value of the measuring device. This means that degassing with vapor bubbles is well worth applying to most degassing processes. All that remains seems to be how to optimize for the requirements of the individual applications.

Industrial applications requiring degassing are truly extensive. Vacuum bubbling is a dramatically effective approach to reducing installation and energy costs when applied directly to the degassing processes used in most industrial settings. The reason for this is that the functional structure of the deaerator for vacuum bubbling is simple, the volume of vapor bubbles is maximized at low pressure, and a high level of degassing is possible due to the oxygen concentration in the bubbles close to zero. An interesting application of degassing, albeit of less direct economic demand, may be mentioned for aircraft fuel. For aircraft fuel, the solubility of oxygen and nitrogen in air is very high compared to water. This dissolved gas is exhausted when flying in a high-altitude, low-pressure atmosphere and can become an element that threatens the safety of the aircraft, so all aircraft are obliged to install a separate safety device to suppress it. However, if the dissolved oxygen in the fuel is pretreated at the end of the manufacturing process or before it is injected into the fuel tank, assuming that there is no significant loss in the combustion characteristics of the fuel, the safety of all air passengers on Earth will be guaranteed in advance. It seems that the required level of deaeration to be applied here can be met by the phase 1 degassing, which can be applied in a relatively low level of vacuum that is higher than the vapor pressure of the liquid. Deterioration issues due to oxidation of biofuels, including SAF, are known to be one of the major obstacles that ban their commercialization. As one of the solutions to this issue, degassing of biofuel is expected to play a meaningful role. This is my personal opinion, but degassing biofuels using other existing degassing processes is probably not possible or economically not feasible. Until now, the development of additives to prevent oxidation has been almost the only approach to control the oxidation of biofuels, but I propose the application of the vacuum bubbling degassing method and related research as a way to control the oxidation of biofuels, including SAF.

The components of deaerated air, which is the result of degassing of water, are governed by the solubility of dissolved gases in the range where the fraction of water vapor does not increase due to the intervention of vapor bubbles. Since the solubility of oxygen in water is about twice that of nitrogen and the partial pressure ratio of oxygen and nitrogen in the atmosphere is about 1:4, if you recall Henry's law, the dissolved gas ratio is 2:4 or 1:2. In other words, if dissolved gas is collected from water under atmospheric pressure, air with an oxygen concentration of about 33% can be obtained, and a system that can utilize this can be expected to emerge or be developed. Efforts to utilize dissolved oxygen in water, represented by "artificial gills," for breathing can contribute to the expansion of human living space, from marine sports (scuba diving) to air-conditioning, various underwater activities, and creation of underwater living spaces.

Finally, the desalination field is not just a technical field but an important issue that is connected to the global quality of life or survival. Before climate change became an issue, desalination was a technology available only to oil-producing countries in the Middle East or countries rich in resources. However, due to the water shortage issue that came along with climate change, it has now become a commonly required technology in most countries. Demand for technology in this field is exploding, but expectations for sustainable technology to meet this growing demand are higher than in any other field. The applicability of vacuum bubbling to desalination cannot yet be determined. Many technologies that started with great ideas, such as desalination technology using renewable energy, have been introduced, but it is still unclear how successful they have been put into practical use. This is because, in the process of technological challenges to achieve successful commercialization, it cannot be considered too thoroughly. Vacuum bubbling will help with deaeration first when applied to phase-change desalination. Then, when applied to desalination itself, there are two challenges to be addressed. The first is the generation of massive vapor, not necessarily in the form of bubble. The amount of room-temperature vapor bubbles obtained through experiments so far is very limited, so it is a % level of input electrical energy. The second is a high-efficiency condenser. Since vapor is in a low-pressure (vacuum) state of about 1 to 5 kPa, the density of vapor should also be very low, so a condenser with high efficiency is essential in a limited temperature difference. Among these, the latter will be realized by the current state-of-the-art technology in this field, but the former is a task to be accomplished by someone's challenge in the future. It could be me or it could be you reading this article. In the process of reviewing the materials while preparing the manuscript, I found out that the room-temperature vapor bubble generation part corresponds to the single-stage flash evaporation process. The applied thermodynamic state is different from the existing one, but the principle is the same. Therefore, if the current top-notch technology in this field is applied to the former part, the time to expect the answer may not be long.

Vacuum bubbling is a bit different from the research direction of previous researchers works which are focused on heating and superheating, which were important in the industrialization process. However, it is just an example of applying a phenomenon that occurs at a little less hot and a little less pressure while using the same academic foundation they have established. I have introduced a few possible applications that have been interesting to me, but the areas where vacuum bubbling can be used will be wider and more diverse. By all means, I hope that the technology in this field that I have introduced can be served as a small step toward creating a better future for humanity and a "sustainable future," as they say these days. Thank you.

Index

Note: Page numbers in *italics* indicate a figure and page numbers in **bold** indicate a table on the corresponding page.

For Product Safety Concerns and Information please contact our EU
representative GPSR@taylorandfrancis.com
Taylor & Francis Verlag GmbH, Kaufingerstraße 24, 80331 München, Germany

www.ingramcontent.com/pod-product-compliance
Lightning Source LLC
Chambersburg PA
CBHW070728220326
41598CB00024BA/3348